THE NATURE *of* TECHNOLOGY

What It Is and How It Evolves

W. BRIAN ARTHUR

FREE PRESS

New York London Toronto Sydney

Free Press
A Division of Simon & Schuster, Inc.
1230 Avenue of the Americas
New York, NY 10020

First Free Press trade paperback edition January 2011

FREE PRESS and colophon are trademarks of Simon & Schuster, Inc.

For information about special discounts for bulk purchases,
please contact Simon & Schuster Special Sales at 1-866-506-1949
or business@simonandschuster.com.

The Simon & Schuster Speakers Bureau can bring authors to your live event.
For more information or to book an event, contact the Simon & Schuster
Speakers Bureau at 1-866-248-3049 or visit our website at
www.simonspeakers.com.

Manufactured in the United States of America

1 3 5 7 9 10 8 6 4 2

Library of Congress Control Number: 2009007015

ISBN 978-1-4165-4405-0
ISBN 978-1-4165-4406-7 (pbk)
ISBN 978-1-4391-6578-2 (ebook)

CONTENTS

THE NATURE *of* TECHNOLOGY

PREFACE

We are often haunted later in life by questions we could not resolve in our teens and twenties. As a rather young undergraduate—I was barely 17 when I started—I was trained as an electrical engineer, and although I could score high on exams, that merely meant I had a knack for mathematics. I was aware somehow I did not really understand the essence of what I was studying: what the real nature of technology was. Our professors told us variously that technology was the application of science; that it was the study of the machinery and methods used in the economy; that it was society's knowledge of industrial processes; that it was engineering practice. But none of these seemed satisfactory. None seemed to get in any way to the "technology-ness" of technology, and I was left with the question unanswered.

Later on in graduate school, when I switched from engineering, I became fascinated with how the economy develops and builds out. It was clear to me that the economy was in no small part generated from its technologies. After all, in a sense an economy was nothing more than the clever organization of technologies to provide what we need. Therefore it would evolve as its technologies evolved. But if that was so, then how did technologies evolve, and where were they generated from? How did economies give birth to their technologies? What precisely was technology anyway? I was back at the same question.

For many years I gave the question little thought. But in the early 1980s my attention was drawn back to technology when I was exploring increasing returns in the economy. Technologies, in the form of new technical products and processes—think of early automobiles—improved with use and adoption, and this led to further use and adoption, creating a positive feedback or increasing returns to adoption. Increasing returns created a problem for economics. If two products (or technologies, in my case) with increasing returns competed, the one that got ahead might get further ahead and hence dominate the market. But the winner could be either one: there were multiple possible outcomes. So how does one of these get selected? The approach I developed allowed that random events, magnified by the inherent positive feedbacks, might over time select the outcome probabilistically. We could analyze increasing returns situations if we saw them as (partly) random processes. The idea worked.

To find good examples, I began in 1981 to look at particular technologies and how they had developed. These bore out my theory. But what caught my attention was something that was not connected directly with increasing returns, something that seemed vague at first. I realized that new technologies were not "inventions" that came from nowhere. All the examples I was looking at were created—constructed, put together, assembled—from previously existing technologies. Technologies in other words consisted of other technologies, they arose as combinations of other technologies. The observation was simple enough and at first did not seem particularly important. But it meant, I realized, that if new technologies were constructed from existing ones, then considered collectively, technology created *itself*. Later I came upon the work of Francisco Varela and Humberto Maturana on self-producing systems, and I could tell people impressively that technology was *autopoietic* (or self-creating). But in the mid-1980s I was not aware of Varela and Maturana. All I could do was stare at this universe of self-creating objects and wonder about the consequences of this self-creation.

Slowly it became clear that combination might be a key to figuring out realistic mechanisms of the invention and evolution of technology, things that had not been satisfactorily treated in the previous thinking about technology. Some mechanisms fell into place in the 1990s—I published the idea of structural deepening in 1994—and I was vaguely aware of others.

I got caught up with other questions in the 1990s, mainly with issues of complexity and cognition in the economy. And only around 2000 did I begin to think systematically again about technology and how it gets generated. I began to realize, again slowly, that other principles besides combination were at work. Technologies consisted of parts—assemblies and subassemblies—that were themselves technologies. So technologies had a recursive structure. And every technology, I realized, was based upon a phenomenon, some effect it exploited, and usually several. Therefore, collectively technology advanced by capturing phenomena and putting them to use. I also began to see that the economy was not so much a container for its technologies, as I had been implicitly taught. The economy arose from its technologies. It arose from the productive methods and legal and organizational arrangements that we use to satisfy our needs. It therefore issued forth from all these capturings of phenomena and subsequent combinations.

Stalking these ideas brought me into the library stacks at Stanford, browsing for writings on technology. At first the volume of work seemed large. But on reflection I realized that it was small, curiously small, considering that the collective of technology is as large and complicated and interesting as the economy or the legal system. There were plenty of writings—and textbooks galore—on particular technologies, especially those to do with fashionable ones such as computation or biotechnology. But on the nature of technology and of innovation and of the subsequent evolution of technologies, I could find little. I found meditations on technology from engineers and French philosophers, studies of the adoption and diffusion of technologies, theories of how society influences technology and technology influences society, and observations on how technologies are designed and subsequently develop. But I was tracking something deeper. I wanted discussion of principles behind technology, some common logic that would structure technology and determine its ways and progress, and I could not find these. There was no overall theory of technology.

I will draw in this book upon the discussions of technology that I could find. These are produced by a small coterie of thinkers that include philosophers, engineers, social scientists, and historians. And all are useful. But those that turned out to be most useful were the detailed and beautiful case studies of the historians on how particular technologies came into being. I puzzled why historians of all people had the most to say about the ways and essence of technology and its innovation. Later it came to me that a great deal more of the world emerges from its technologies than from its wars and treaties, and historians are naturally interested in how the world has formed itself. They are therefore interested in how technologies come into being.

This book is an argument about what technology is and how it evolves. It grew out of two sets of lectures I gave: the 1998 Stanislaw Ulam Memorial Lectures at the Santa Fe Institute on "Digitization and the Economy"; and the Cairnes Lectures in 2000 at the National University of Ireland, Galway, on "High Technology and the Economy." It uses material from both series, but builds largely from the Cairnes ones.

I have had to make some decisions in the writing of this book. For one, I decided to write it in plain English (or what I hope is plain English). I am a theorist by profession and nature, so I have to admit this has caused me some horror. Writing a book about serious ideas for the general reader was common a hundred or more years ago, but today it lays you open to a charge of nonseriousness. Certainly in the fields I know best—economics and engineering—you signal your prowess by expressing theories in specialist and arcane terms (something I have done plenty of in the past).

Still, the idea of a book that would be both serious and accessible to the general reader won out for several reasons. One is that simple honesty required it: the subject has not been thought out in great detail before, so no specialist arcana are required. Another is that I believe technology is too important to be left to specialists and that the general reader needs to be brought in. And not least, I want to stir wide interest in what I believe is a subject of great beauty, one that I believe has a natural logic behind it.

Early on I found out that words were a problem in technology. Many of the ones used most heavily—"technology" itself, "innovation," "technique"—have overlapping and often contradicting meanings. "Technology" has at least half-a-dozen major meanings, and several of these conflict. Other words were loaded with emotional associations. "Invention" conjures up pictures of lone inventors struggling against odds, which feeds the popular notion that novel technologies spring from the brows of extraordinary geniuses and not from any particular ancestry. I began to realize that many of the difficulties in thinking about technology stemmed from imprecision in how words are used. So, as the argument developed I found myself working almost as a mathematician would, precisely defining terms and then drawing out the consequences and properties that logically derive from these. The result, as the reader will see, is a constant—and necessary—concern with words and how they are used in technology. Also necessary was the introduction of a few new terms. I was hoping to avoid this, but some are needed for the argument and have slipped in.

I also had to decide about examples, and chose them from a wide range. Sticking with a narrow set of examples to illustrate what I have to say might have been more convenient. One prospective publisher suggested I stick with something simple, such as the pencil. But my point is that there exists a logic to technology that applies across all sorts of instances: to computer algorithms and beer brewing, power stations and pencils, handheld devices and DNA sequencing techniques. So the examples range all over. To make them intelligible and save unnecessary explaining, I have tried to choose technologies the reader is likely to be familiar with.

I want to say a word or two about what this book is not. It is not about the promises and threats of technology for the future of society and the environment. These are important, but are not part of my argument. It is not about particular technologies, nor about the new technologies that are coming down the pipeline. It is not a comprehensive overview of the engineering process; that has been amply covered before. And it is not a discussion of the human side of creating technology. People are required at every step of the processes that create technology; but I concentrate on the logic that drives these processes, not on the humans involved in them. I decided early on to discuss topics only if they bore directly on my argument. So several worthy topics are not mentioned except in passing: the sociology of invention, the adoption and diffusion of technology, the ideas of cost-push and demand-pull, the role of institutions and of learned societies, the history of technology. These are all important, but are not covered here.

This book is also not a review of the literature on technology, though it will certainly touch on the work of others. I often think of investigations such as this as Lewis and Clark expeditions. These start from well-known territory, quickly come into new land, then here and there happen upon previously occupied terrain. This exploration is no exception. We will come upon travelers who have been here before—in this territory Heidegger has left traces and Schumpeter's footprints are everywhere. I will mention these as we encounter them. But I leave discussion of the previous literature mainly to the notes. To explorers who have traveled in this terrain whom I have missed, I offer apologies.

One last disclaimer. Because I write a book on technology, the reader should not take it that I am particularly in favor of technology.

Preface

Oncologists may write about cancer, but that does not mean they wish it upon people. I am skeptical about technology, and about its consequences. But I have to admit some things. I have a passion for science and I am enthralled by the magic of technology. And I confess a fondness for aircraft. And for old-fashioned radio electronics.

W. Brian Arthur
Santa Fe Institute, New Mexico;
and Intelligent Systems Laboratory,
PARC, Palo Alto, California.

1

QUESTIONS

I have many attitudes to technology. I use it and take it for granted. I enjoy it and occasionally am frustrated by it. And I am vaguely suspicious of what it is doing to our lives. But I am also caught up by a wonderment at technology, a wonderment at what we humans have created. Recently researchers at the University of Pittsburgh developed a technology that allows a monkey with tiny electrodes implanted in its brain to control a mechanical arm. The monkey does this not by twitching or blinking or making any slight movement, but by using its thoughts alone.

The workings behind this technology are not enormously complicated. They consist of standard parts from the electronics and robotics repertoires: circuits that detect the monkey's brain signals, processors and mechanical actuators that translate these into mechanical motions, other circuits that feed back a sense of touch to the monkey's brain. The real accomplishment has been to understand the neural circuits that "intend" motion, and tap into these appropriately so that the monkey can use these circuits to move the arm. The technology has obvious promise for impaired people. But that is not what causes me wonder. I wonder that we can put together circuits and mechanical linkages—in the end, pieces of silicon and copper wiring, strips of metal and small gears—so that machinery moves in response to thought and to thought alone.

I wonder at other things we can do. We put together pieces of metal alloy and fossil fuel so that we hurtle through the sky at close to the speed of sound; we organize tiny signals from the spins of atomic nuclei to make images of the neural circuits inside our brains; we organize biological objects—enzymes—to snip tiny slivers of molecules from DNA and paste them into bacterial cells. Two or three centuries ago we could not have imagined these powers. And I find them, and how we have come by them, a wonder.

Most of us do not stop to ponder technology. It is something we find useful but that fades to the background of our world. Yet—and this is another source of wonder for me—this thing that fades to the background of our world also creates that world. It creates the realm our lives inhabit. If you woke some morning and found that by some odd magic the technologies that have appeared in the last six hundred years had suddenly vanished: if you found that your toilet and stove and computer and automobile had disappeared, and along with these, steel and concrete buildings, mass production, public hygiene, the steam engine, modern agriculture, the joint stock company, and the printing press, you would find that our modern world had also disappeared. You—or we, if this strange happening befell all of us—would still be left with our ideas and culture, and with our children and spouses. And we would still have technologies. We would have water mills, and foundries, and oxcarts; and coarse linens, and hooded cloaks, and sophisticated techniques for building cathedrals. But we would once again be medieval.

Technology is what separates us from the Middle Ages; indeed it is what separates us from the way we lived 50,000 or more years ago. More than anything else technology creates our world. It creates our wealth, our economy, our very way of being.

What then is this thing of such importance? What is technology in its nature, in its deepest essence? Where does it come from? And how does it evolve?

These are the questions I will ask in this book.

———

Maybe we can simply accept technology and not concern ourselves much with the deeper questions behind it. But I believe—in fact I believe fervently—that it is important to understand what technology is and how it comes to be. This is not just because technology creates much of our world. It is because technology at this stage in our history weighs on us, weighs on our concerns, whether we pay attention to it or not. Certainly technology has enabled our children to survive where formerly they might have died; it has prolonged our own lives and made them a great deal more comfortable than those of our ancestors just two or three centuries ago; it has brought us prosperity. But it has also brought us a profound unease.

This unease does not just come from a fear that technologies cause new problems for every problem they solve. It wells up also from a deeper and more unconscious source. We place our hopes in technology. We hope in technology to make our lives better, to solve our problems, to get us out of predicaments, to provide the future we want for ourselves and our children. Yet, as humans, we are attuned not to this thing we hope in—not to technology—but to something different. We are attuned in the deepest parts of our being to nature, to our original surroundings and our original condition as humankind. We have a familiarity with nature, a reliance on it that comes from three million years of at-homeness with it. We trust nature. When we happen upon a technology such as stemcell regenerative therapy, we experience hope. But we also immediately ask how natural this technology is. And so we are caught between two huge and unconscious forces: Our deepest hope as humans lies in technology; but our deepest trust lies in nature. These forces are like tectonic plates grinding inexorably into each other in one long, slow collision.

The collision is not new, but more than anything else it is defining our era. Technology is steadily creating the dominant issues and upheavals of our time. We are moving from an era where machines enhanced the natural—speeded our movements, saved our sweat, stitched our clothing—to one that brings in technologies that resemble or replace the natural—genetic engineering, artificial intelligence, medical devices implanted in our bodies. As we learn to use these technologies, we are moving from using nature to intervening directly within nature. And so the story of this century will be about the clash between what technology offers and what we feel comfortable with. No one claims that the nature and workings of technology are simple; there is no reason to think they are simpler than the nature and workings of the economy or of the law. But they are determining for our future and our anxieties about it.

This book is not about the benefits or evils of technology, there are other books that look at these. It is an attempt to understand this thing that creates so much of our world and causes us so much unconscious unease.

And this brings us back to the same questions. What is technology? What is it in the deepest sense of its nature? What are its properties and principles? Where does it come from—how does it come into being? How does it develop? And how does it evolve?

Missing: An "-ology" of Technology

One good place to start is to ask what we really know about technology. The reader might expect the answer here to be straightforward, but actually it is not. In fact it is almost paradoxical: we know a great deal about technology and we know little. We know a great deal about technologies in their individual sense, but much less about technology in the way of general understandings. We know all the particularities about the individual methods and practices and machinery we use—or at least some people, the designers of these, do. We know every step in the production of a computer microprocessor, and every part of the processor, and every part of every part. We know exactly how the processor operates, and all the pathways of the electrons inside it. And we know how the processor fits with the other components in a computer, how it interfaces with the BIOS chip and the interrupt controller. We know exactly these things—exactly what lies within each technology—because we have placed them all there in all their detail. Technology in fact is one of the most completely known parts of the human experience. Yet of its essence—the deep nature of its being—we know little.

This sort of contrast between known content and less-known principles is not rare. Around two centuries ago, in the time of the French zoologist Georges Cuvier, biology (or natural history as it was then called) was a vast body of knowledge on individual species and their comparative anatomy and their interrelations. "Today," he said, writing in 1798, "comparative anatomy has reached such a point of perfection that, after inspecting a single bone, one can often determine the class, and sometimes even the genus of the animal to which it belonged." Cuvier was only slightly exaggerating. Naturalists did have detailed knowledge, and they were deeply aware of the family relationships among animals. But they had few principles to hold all this knowledge together. They had no clear idea of how animals had come to be; no mechanism by which evolution—if it existed—could work; no obvious idea of whether animals could modify their parts or how this could happen. All this came later, as principles were found.

We are in the same position with technology. We have detailed studies about the history of individual technologies and how they came into being. We have analyses of the design process; excellent work on how economic factors influence the design of technologies, how the adoption process works, and how technologies diffuse in the economy. We have analyses of how society shapes technology, and of how technology shapes society. And we have meditations on the meaning of technology, and on technology as determining—or not determining—human history. But we have no agreement on what the word "technology" means, no overall theory of how technologies come into being, no deep understanding of what "innovation" consists of, and no theory of evolution for technology. Missing is a set of overall principles that would give the subject a logical structure, the sort of structure that would help fill these gaps.

Missing, in other words, is a theory of technology—an "-ology" of technology.

There is no clear reason why this is so. But I strongly suspect that because technology stands in the shadow of its more prestigious sister, science, we honor it less—and therefore study it less. And I suspect that because we feel technology to be the cause of much disharmony in our world, at some unconscious level we feel it to be intellectually distasteful—unworthy perhaps of deep study. We also feel vaguely that because we have created technology, we already understand it.

And there is another reason. The people who have thought hardest about the general questions of technology have mostly been social scientists and philosophers, and understandably they have tended to view technology from the outside as stand-alone objects. There is the steam engine, the railroad, the Bessemer process, the dynamo, and each of these is a boxed-up object with no visible insides—a black box, to use economic historian Nathan Rosenberg's term. Seeing technologies this way, from the outside, works well enough if we want to know how technologies enter the economy and spread within it. But it does not work well for the fundamental questions that interest us. It is like viewing the animal kingdom as a collection of separate black-boxed species: lemurs, macaques, zebras, platypuses. With no obvious relation between these and no inside anatomies to compare, it would be difficult to see how they are related as species, how they originate in the first place, and how they subsequently evolve. So it is with technologies. If we want to know how they relate to each other, and how they originate and subsequently evolve, we need to open them up and look at their inside anatomies.

I want to be fair in what I say here. Social scientists are aware that technologies have inside components, and in many cases are well aware of how these work together to produce the technology. And some historians have opened up technologies to look in detail at how they originated and changed over time. But most of this "inside thinking" concerns itself with particular technologies—radio, radar, the Internet—and not with technologies in a general sense. Things might have been different if engineers had been the main thinkers about technology; they naturally see technologies from the inside. I once asked the distinguished technologist Walter Vincenti why so few engineers had attempted to lay down a theoretical foundation for their field. "Engineers," he told me, "like problems they can solve."

Evolution in Technology

One of the problems *I* want to solve, certainly one of the deeper questions about technology, is how it evolves. Or, I should say, *whether* it evolves, because it is not clear without argument that technology evolves at all. The word "evolution" has two general meanings. One is the gradual development of something, as with the "evolution" of ballet or the English madrigal. I will call this evolution in the narrow sense, or more usually "development." The other is the process by which all objects of some class are related by ties of common descent from the collection of earlier objects. This is evolution in its full sense, and it is what I will mean by evolution.

For me, how technology evolves is the central question in technology. Why do I believe this? Without evolution—without a sense of common relatedness—technologies seem to be born independently and improve independently. Each must come from some unexplained mental process, some form of "creativity" or "thinking outside the box" that brings it into existence and separately develops it. With evolution (if we can find how it works), new technologies would be birthed in some precise way from previous ones, albeit with considerable mental midwifing, and develop through some understood process of adaptation. In other words, if we could understand evolution, we could understand that most mysterious of processes: innovation.

The idea of evolution in technology is by no means new. Barely four years after Darwin's *Origin of Species*, Samuel Butler was calling for a theory of the "mechanical kingdom" that would explain "that part among machines which natural selection has performed in the animal and vegetable kingdoms. . . ." His essay, "Darwin Among the Machines," is full of the enthusiasms of the time: "[t]here is nothing which our infatuated race would desire more than to see a fertile union between two steam engines; it is true that machinery is even at this present time employed in begetting machinery, in becoming the parent of machines often after its own kind, but the days of flirtation, courtship, and matrimony appear to be very remote." This, of course, is hyperbole. Still, if I take the essay seriously, I cannot avoid feeling Butler is trying to shoehorn technology into a framework—Darwin's biological evolution—that might not be appropriate for it.

What *is* clear from the historical record is that modern versions of certain particular technologies do descend from earlier forms. About seventy years after Butler, the sociologist S. Colum Gilfillan traced the descent of the ship from the dugout canoe to the sailing ship to the modern steamship of his day. Gilfillan was a member of a small American school of historians and sociologists that was deeply interested in technology and invention in the 1920s and 30s. And he was knowledgeable about ships; he had been curator of shipping for Chicago's Museum of Science and Industry. In 1935 he traced in historical detail how each of the "inventions" of planking, ribbing, fastenings, keels, lateen sails, and square sails came about (he devotes four pages to the origins of the gaff sail alone); how these gradually transformed the most primitive floating objects into the sailing ship; and how further inventions metamorphosed the sailing ship into the modern steamship. This is not evolution in its full sense. It is "evolution" in the narrow sense of gradual development: the descent of form. What Gilfillan showed is that for some technologies, certainly for ships, we can trace a detailed line of descent.

For a full theory of evolution we would need something more. We would need an argument that all technologies, not just some, are descended from earlier technologies, and an explicit mechanism by which this happens. Attempts to provide such an argument have been few, and unsuccessful. Most, like Butler's, have been more proposals than theories, and all base their reasoning squarely on Darwin's. The central idea works this way. A given technology, the railroad locomotive, say, exists at a particular time in many variants. This is because it has different purposes to fulfill, different environments to operate in (different "habitats" to adapt to, if you like), and different designers who use different ideas. From these variations, some perform better and are selected for further use and development; they pass on their small differences to future designs. We can then follow Darwin and say that "it is the steady accumulation, through natural selection, of such differences, when beneficial to the individual, that gives rise to all the more important modifications of structure." In this way technology evolves.

The argument sounds reasonable, but it quickly runs into a difficulty. Some technologies—the laser, the jet engine, radar, the Quicksort computer algorithm, the railroad locomotive itself—just appear, or at least they seem to just appear, and unlike novel biological species, they are not versions of earlier objects. The jet engine is not a variation of the internal combustion engine or anything else that preceded it, and it did not come into being by the steady accumulation of small changes in its predecessors. So explaining "novelty," meaning abrupt radical novelty, becomes a major obstacle for technology evolutionists. The appearance of radically novel technologies—the equivalent of novel species in biology—cannot be accounted for.

One way out, a rather extreme one, is to lean harder on Darwin and say that if different designers bring forth different variants of a technology, some of these variants and the ideas behind them may be radical. Change can therefore be radical and abrupt, as well as gradual. This sounds plausible enough, but if you look into what it would require in practice for any radical innovation it does not stand up. Radar "descends" from radio. But you can vary 1930s radio circuits all you like and as radically as you like and you will never get radar. Even if you vary ideas about radio circuits all you like you will still not get radar. Radar requires a different principle than radio.

I do not want to dismiss variation and selection in technology. Certainly technologies exist in multiple versions and certainly superior performers are selected, so that later forms can indeed descend in this way from earlier ones. But when we face the key question of how radically novel technologies originate—the equivalent of Darwin's question of how novel species originate in biology—we get stymied. Darwin's mechanism does not work.

Combinatorial Evolution

There is a way toward understanding how technology evolves, but to get there we need to shift our thinking. What we should really be looking for is not how Darwin's mechanism should work to produce radical novelty in technology, but how "heredity" might work in technology. If evolution in its fullest sense holds in technology, then all technologies, including novel ones, must descend in some way from the technologies that preceded them. This means they must link with—be "sired" by—certain technologies that preceded them. In other words, evolution requires a mechanism of "heredity," some detailed connection that links the present to the past. From the outside it is impossible to see this mechanism—looked at as a blackboxed device, it is hard to say how the laser has come into being from previous technologies.

What if we looked inside technologies? Would we see anything that would tell us how novelty works in technology? Would we see anything that could yield a proper theory of evolution for technology?

If you open up a jet engine (or aircraft gas turbine powerplant, to give it its professional name), you find components inside—compressors, turbines, combustion systems. If you open up other technologies that existed before it, you find some of the same components. Inside electrical power generating systems of the early twentieth century were turbines and combustion systems; inside industrial blower units of the same period were compressors. Technologies inherit parts from the technologies that preceded them, so putting such parts together—combining them—must have a great deal to do with how technologies come into being. This makes the abrupt appearance of radically novel technologies suddenly seem much less abrupt. Technologies somehow must come into being as fresh combinations of what already exists.

So far, this is only a hint of something we can use to explain novelty. But built up properly it will be central to my argument. Novel technologies must somehow arise by combination of existing technologies.

Actually, this idea, like evolution itself, is by no means new. It has been mooted about by various people for well over 100 years, among them the Austrian economist Joseph Schumpeter. In 1910 Schumpeter was twenty-seven, and he was concerned not directly with combination and technology but with combination in the economy. "To produce," he said, "means to combine materials and forces within our reach. . . . To produce other things, or the same things by a different method, means to combine these materials and forces differently." Change in the economy arose from "new combinations of productive means." In modern language we would say it arose from new combinations of technology.

Schumpeter had come to this idea because he had been asking a seemingly simple question: how does an economy develop? (In modern language we would say, how does it change structurally?) External factors of course can change an economy: if it discovers a new source of raw materials, or starts to trade with a new foreign partner, or opens up new territories, its structure can change. But Schumpeter was asking whether an economy could change itself without external factors—purely from within—and if so how. The prevailing doctrine of the time, equilibrium economics, held that it could not. Without outside disturbances, the economy would settle into a static pattern or equilibrium, fluctuate around this, and stay there. Schumpeter, however, realized there was "a source of energy within the economic system which would of itself disrupt any equilibrium that might be attained." That source was combination. The economy continually created the new by combining the old, and in doing so it disrupted itself constantly from within.

Schumpeter's book did not appear in English until 1934, and by then other people in the 1920s and 30s had come to the same conclusion: combination drove change—or at least the innovation of technology. Invention, said historian Abbott Payson Usher in 1929—another member of the American school—proceeds from "the constructive assimilation of pre-existing elements into new syntheses." Gilfillan himself put it more succinctly: an invention is a "new combination of prior art." After that the idea drifted, occasionally mentioned but not much invoked, in part because nobody—not Schumpeter, not Usher, not Gilfillan, nor anyone else—explained how such combination could bring forth a new invention. It is easy enough to say that the jet engine is a combination of parts available to its inventors, Frank Whittle and Hans von Ohain, but not easy to explain how such combination takes place in the minds of a Whittle or von Ohain.

Combination at least suggests a way by which novelty arises in technology. But this merely links individual novel technologies back to particular technologies that existed before. It does not yet give us a sense of the whole of technology building up from what went before. For that we need to add a second piece to the argument. If new technologies are indeed combinations of previous ones, then the stock of existing technologies must somehow provide the parts for combination. So the very cumulation of earlier technologies begets further cumulation.

This idea has a history too. One of Schumpeter's near contemporaries, the American William Fielding Ogburn, pointed this out in 1922. Ogburn was a sociologist, and again very much a member of the American school. He was fascinated by what generated social change (or in his language, change in material culture). And like Schumpeter he saw the combination of previous technologies—invention—as the source of change. But he also saw something else: that inventions built cumulatively from earlier inventions. "It would seem that the larger the equipment of material culture the greater the number of inventions. The more there is to invent with, the greater will be the number of inventions." This explained why more "primitive" societies could not invent our modern technologies; they did not possess the necessary ingredients and knowledge of how to work with them. "The street car could not have been invented from the material culture existing at the last glacial period. The discovery of the power of steam and the mechanical technology existing at the time made possible a large number of inventions." The insight here is marvelous. But sadly, it ends there. Ogburn does not use it to construct any theory of technology or of its evolution, which he easily could have.

If we put these two pieces together, that novel technologies arise by combination of existing technologies and that (therefore) existing technologies beget further technologies, can we arrive at a mechanism for the evolution of technology? My answer is yes. Stated in a few words this would work as follows. Early technologies form using existing primitive technologies as components. These new technologies in time become possible components—building blocks—for the construction of further new technologies. Some of these in turn go on to become possible building blocks for the creation of yet newer technologies. In this way, slowly over time, many technologies form from an initial few, and more complex ones form using simpler ones as components. The overall collection of technologies bootstraps itself upward from the few to the many and from the simple to the complex. We can say that technology creates itself out of itself.

I will call this mechanism evolution by combination, or more succinctly, *combinatorial evolution*.

Of course, the argument as I have stated it is not quite complete. Combination cannot be the only mechanism behind technology's evolution. If it were, modern technologies such as radar or magnetic resonance imaging (the MRI of hospital use) would be created out of bow-drills and pottery-firing techniques, or whatever else we deem to have existed at the start of technological time. And we would have a problem designating the start of this technological time. If bow-drills and pottery-firing techniques themselves formed by combination of earlier technologies, where did these ur-technologies come from? We land in an infinite regress. Something else, something more than mere combination, must be going on to create novel technologies.

That something else, I will argue, is the constant capture of new natural phenomena and the harnessing of these for particular purposes. In the cases of radar and MRI, the harnessed phenomena are the reflection of electromagnetic waves and nuclear magnetic resonance, and the purposes are the detection of aircraft and diagnostic imaging of the body. Technology builds out not just from combination of what exists already but from the constant capturing and harnessing of natural phenomena. At the very start of technological time, we directly picked up and used phenomena: the heat of fire, the sharpness of flaked obsidian, the momentum of stone in motion. All that we have achieved since comes from harnessing these and other phenomena, and combining the pieces that result.

In bare-bones form like this the argument is easy to state, but to make it precise many details will need to be worked out. I will have to specify what it really means that novel technologies are "combinations" of existing ones. Technologies are not thrown together randomly as combinations of existing components, so I will have to provide the detailed mechanics of how combination works—of how, say, the turbojet arises as a combination of existing things. This means, to take things back a further step, we will have to look at how technologies are logically structured, because combination—however it happens—must take place in accordance with that structure. We will have to look at the considerable part human beings, in particular their minds, play in this combination process; new technologies are constructed mentally before they are constructed physically, and this mental process will need to be carefully looked into. We will have to pay attention to why technologies come into existence at all: how human needs call for the creation of new technologies. We will have to make clear what it means that technologies beget further technologies—that novel technologies issue forth from the collective of existing ones. And to take things right back to fundamentals, we will have to define clearly what we mean by "technology."

The Themes of the Book

This book is an argument about what technology is and how it evolves. It is an attempt to construct a theory of technology, "a coherent group of general propositions," we can use to explain technology's behavior. In particular it is an attempt to create a theory of evolution for technology.

My plan is to start from a completely blank state, taking nothing about technology for granted. I will build the argument piece by piece from three fundamental principles. The first will be the one I have been talking about: that technologies, all technologies, are combinations. This simply means that individual technologies are constructed or put together—combined—from components or assemblies or subsystems at hand. The second will be that each component of technology is itself in miniature a technology. This sounds odd and I will have to justify it, but for now think of it as meaning that because components carry out specific purposes just as overall technologies do, they too qualify as technologies. And the third fundamental principle will be that all technologies harness and exploit some effect or phenomenon, usually several.

I will say more about these central principles as we go. But notice they immediately give us a view of technologies from the inside. If technologies are combinations they immediately have an interior: they are assembled from parts and groups of parts to meet their purpose. And this interior consists of parts and subsystems that themselves are technologies. We can begin to see that novel technologies originate by piecing together existing ones, and of course by capturing phenomena. We can see technologies developing by changing these interior parts, by swapping in better ones that improve their performance. And we can see different technologies as possessing internal parts inherited in common from previous technologies. Viewed this way technology begins to acquire a "genetics." Nothing equivalent to DNA or its cellular workings of course, or as beautifully ordered as this. But still, a rich interlinked ancestry.

All this sounds organic—very organic—and indeed the view we will be led to is as much biological as mechanical. For sure, technologies are not biological organisms, they are mechanistic almost by definition: whether they are sorting algorithms or atomic clocks, they have parts that interact in predictable ways. But once we lay them out as combinations that can be combined into further combinations, we begin to see them not so much as individual pieces of clockwork but as complexes of working processes that interact with other complexes to form new ones. We see a world where the overall collective body of technology forms new elements—new technologies—from existing ones. Technology builds itself organically from itself, and this will be one of the themes of this book.

The change in vision I am proposing, from seeing technologies as stand-alone objects each with a fixed purpose to seeing them as objects that can be formed into endless new combinations, is not just abstract. It is mirrored in a broad shift in the character of technology currently taking place. The old industrial-process technologies of the sort that characterized the manufacturing economy—the open-hearth process for steelmaking, the cracking process for refining crude oil—were indeed for the most part fixed. They did one thing only and did it in a fixed place: they processed particular raw material inputs into particular industrial outputs and did this largely in separate, stand-alone factories. But now these relatively independent processing technologies are giving way to technologies of a different type. These can be easily combined and they form building blocks that can be used again and again. Global positioning technology provides direct location, but it rarely stands alone. It is used as an element in combination with other elements to navigate aircraft and ships, to help survey territory, to manage agriculture. It is like a highly reactive building block in chemistry—the hydroxyl ion, say—doing little on its own, but appearing in a host of different combinations. The same can be said for other elements of the digital revolution: algorithms, switches, routers, repeaters, web services. And we can say the same for the elements that comprise modern genetic engineering or nanotechnology. These can be fitted together in endless combinations that can be configured and reconfigured on the fly for different purposes as conditions demand. These too form building blocks available for continual combination.

Modern technology is not just a collection of more or less independent means of production. Rather it is becoming an open language for the creation of structures and functions in the economy. Slowly, at a pace measured in decades, we are shifting from technologies that produced fixed physical outputs to technologies whose main character is that they can be combined and configured endlessly for fresh purposes.

Technology, once a means of production, is becoming a chemistry.

In attempting to come up with a theory of technology, our first challenge will be to see if we can say something general about it. It is not clear a priori that we can. Hydroelectric power, the process of plastic injection molding, and beekeeping, to choose three technologies at random, seem to have nothing in common. But we will see in the next chapter that technologies do share a common logic in the way they are put together. And this will tell us much about how combination has to work, how technologies come into being, how they subsequently develop, and how they evolve.

But before we get to that we have to settle an even more fundamental question. What exactly is technology, anyway?

2

COMBINATION AND STRUCTURE

What are we talking about when we speak of "technology?" What *is* technology?

The answer, whether we turn to dictionaries or to the writings of technology thinkers, is annoyingly unclear. Technology, we are told, is a branch of knowledge, or the application of science, or a study of techniques, or a practice, or even an activity. The *Oxford English Dictionary* declares with a lovely stuffiness that technology is "the collection of mechanical arts that are available to a culture to make its economy and society function." Presumably the "mechanical arts" are the methods, practices, and devices a culture uses to make things function.

All this may actually be fine; often words carry multiple meanings. But if we accept this, can technology really be knowledge and applied science and a study of something and a practice and a collection? All at the same time? Definitions matter because how we think of technology will determine how we think of it coming into being. If technology is knowledge, it must originate in some way as knowledge; if it is a practice, it must arise through practice; if applied science, it must derive somehow from science. If these definitions are an indication of what we understand about technology, then at the very least they are badly fused together and possibly even contradictory.

We need to cut through the muddle. To do this I will go back to first principles and define technology from scratch.

I will give technology three definitions that we will use throughout the book.

The first and most basic one is that a technology is *a means to fulfill a human purpose*. For some technologies—oil refining—the purpose is explicit. For others—the computer—the purpose may be hazy, multiple, and changing. As a means, a technology may be a method or process or device: a particular speech recognition algorithm, or a filtration process in chemical engineering, or a diesel engine. It may be simple: a roller bearing. Or it may be complicated: a wavelength division multiplexer. It may be material: an electrical generator. Or it may be nonmaterial: a digital compression algorithm. Whichever it is, it is always a means to carry out a human purpose.

The second definition I will allow is a plural one: technology as an *assemblage of practices and components*. This covers technologies such as electronics or biotechnology that are collections or toolboxes of individual technologies and practices. Strictly speaking, we should call these bodies of technology. But this plural usage is widespread, so I will allow it here.

I will also allow a third meaning. This is technology as the entire *collection of devices and engineering practices available to a culture*. Here we are back to the *Oxford's* collection of mechanical arts, or as *Webster's* puts it, "the totality of the means employed by a people to provide itself with the objects of material culture." We use this collective meaning when we blame "technology" for speeding up our lives, or talk of "technology" as a hope for mankind. Sometimes this meaning shades off into technology as a collective activity, as in "technology is what Silicon Valley is all about." I will allow this too as a variant of technology's collective meaning. The technology thinker Kevin Kelly calls this totality the "technium," and I like this word. But in this book I prefer to simply use "technology" for this because that reflects common use.

The reason we need three meanings is that each points to technology in a different sense, a different category, from the others. Each category comes into being differently and evolves differently. A technology-singular—the steam engine—originates as a new concept and develops by modifying its internal parts. A technology-plural—electronics—comes into being by building around certain phenomena and components and develops by changing its parts and practices. And technology-general, the whole collection of all technologies that have ever existed past and present, originates from the use of natural phenomena and builds up organically with new elements forming by combination from old ones.

I will have more to say about these second and third categories of technology in the chapters ahead, particularly on how the collective of technology evolves. But because technologies-singular—individual technologies—make up this collective, I want to focus on them for the rest of this chapter. We need to get clear on what exactly they are and what common logic they share.

A technology, I said, is a means to fulfill a purpose: a device, or method, or process. A technology does something. It executes a purpose. To emphasize this I will sometimes talk of a technology as an *executable*. Here we land in a little trouble. It is easy enough to see a riveting machine as an executable: it is directly activated to carry out a specific task. But what about technologies that we do not think of as being "activated?" What about a bridge? Is a bridge an executable? Or a dam? My answer is that each of these has an ongoing task or set of tasks to carry out. A bridge carries traffic; and a dam stores water or supplies energy. Each of these, if it does not fail, works. Each in this sense executes, and each therefore is an executable.

There is another word I will use frequently throughout: a technology supplies a *functionality*. This is simply the generic task it carries out. GPS (a global positioning system) *locates*—that is its functionality. GPS has many specific purposes: in aircraft navigation, ground location, and surveying. But when we need the generic purpose of locating, GPS is a means that supplies that functionality.

Very well then. But our definition of a technology is still untidy. Means to purposes may be devices or methods or processes and these seem very unalike, so we seem to be talking about very different things. Is this indeed the case? Certainly methods and processes both transform something by a series of stages or steps, so we can lump these together as logically similar. But devices and processes—radio receivers and oil refining, say—seem to be different things. A device seems to be a piece of hardware and not at all like a process. But this is just appearance. A device always processes some thing; it works on that thing from beginning to end to complete the needed task. An aircraft "processes" passengers or cargo from one location to another, and a hammer—if we want to push the idea—"processes" a nail.

A radio receiver processes too. It picks up a radio signal and transforms this into tiny voltage differences in an input antenna. It then uses a resonant circuit to extract a particular frequency from the signal (corresponding to a particular radio station, say); passes the result through a series of amplification stages; then separates out the voice or music (acoustic) information, amplifies the result again, and feeds this output to a loudspeaker or headphones. The radio processes the signal, and I mean this literally, not metaphorically. It pulls signals from the air, purifies them, and transforms them into sounds. It is a miniature extraction process, a tiny factory that these days can be held in the palm of a hand. *All* devices in fact process something. That, after all, is why economists refer to technologies as means of production.

Does the correspondence work in the other direction? Can we view methods and processes as devices? The answer is yes. Processes and methods—think of oil refining or sorting algorithms—are sequences of operations. But to execute, they always require some hardware, some sort of physical equipment that carries out the operations. We could see this physical equipment as forming a "device" executing this sequence of operations. In the case of oil refining this would be a pretty big device. But the point is valid. Processes are devices if we include the equipment that executes them. In this sense oil refining is not different from a radio receiver. Both "process," but one uses large-scale engineering equipment and the other small-scale electronic components.

There is a more general way to see that devices and processes are not different categories. A technology embodies a sequence of operations; we can call this its "software." And these operations require physical equipment to execute them; we can call this the technology's "hardware." If we emphasize the "software" we see a process or method. If we emphasize the "hardware," we see a physical device. Technologies consist of both, but emphasizing one over the other makes them seem to belong to two different categories: devices and processes. The two categories are merely different ways of viewing a technology.

One final thought—more of a question really. When we speak of *a technology*—the cable-stayed bridge, say—are we speaking of it as an instance of a specific device or method (the Pont de Normandie bridge in France, say), or as the *idea* of that device or method (the concept of the cable-stayed bridge)? We can answer this if we realize this question occurs everywhere we abstract and label things. If we talk about a black-bellied plover, sometimes we mean an actual bird—that one running along the beach over there just above the water line; and sometimes we mean a category of bird, the concept of a species of bird (in this case *pluvialis squatarola*). In these cases it is perfectly normal to switch between physical instance and concept as context demands.

We can do the same with technology. We can talk of the Haber process in a particular ammonia factory near Krefeld in Germany, or *the* Haber process as the idea or concept of that technology, and we can switch back and forth between the particular and abstract as needed. This yields a bonus. If we accept something as an abstract concept, we can easily zoom in and out conceptually. We can talk of "the Boeing 787" as a technology, "the passenger aircraft" as a more general technology, and "the aircraft" as an even more general technology. Strictly speaking of course "the aircraft" does not exist. But we find the abstract concept useful because aircraft have certain common parts and architectures we can expect and talk about.

How Technologies Are Structured

I want now to come back to the question I posed at the end of the last chapter. Do technologies—individual ones—share any common logic? In particular, do they share any common structure in the way they are organized?

The answer, as I said, does not look promising. Nevertheless, we shall see shortly that technologies do share a common anatomical structure. Technologies in this regard are like some category of animals—vertebrates, say. Vertebrates differ widely in their anatomical plans and outside appearance; a hippopotamus looks nothing like a snake. But all share the structure of a segmented spinal column, and organs such as chambered hearts, livers, paired kidneys, and a nervous system. All have bilateral symmetry, and all are constructed ultimately from cells. The shared structure of technologies will not be that they contain the same organs. But whatever it is, it will be central to our argument because it constrains how technologies are put together and come into being, much as the structure of haikus (a poem that must have seventeen syllables in three lines of five, seven, and five syllables) constrains how these are put together and come into being.

What then is this common anatomical structure?

To start with, we can say that technologies are put together or combined from component parts or assemblies. A technology therefore is a combination of components to some purpose. This *combination principle* is the first of the three principles of technology I mentioned in the last chapter. A hydroelectric power generator combines several main components: a reservoir to store water, an intake system with control gates and intake pipes called penstocks, turbines driven by the high-energy water flow, electricity generators driven by the turbines, transformers to convert the power output to higher voltage, and an outflow system, or tailrace, to discharge the water. Such assemblies, or subsystems, or subtechnologies (or stages, in the case of process technologies) are groups of components that are largely self-contained and largely partitioned off from other assemblies. Of course some very basic technologies—a rivet, for example—may have only one part. We can allow such elemental technologies as "combinations" in the same way that mathematicians allow sets with only one member. (To be properly mathematical, we should allow for technologies that have zero components and no purpose. No doubt such have been thought of.)

Our first piece of structure, then, is that technologies consist of parts. We can see more structure if we observe that a technology is always organized around a central *concept* or *principle*: "the method of the thing," or essential idea that allows it to work. The principle of a clock is to count the beats of some stable frequency. The principle of radar—its essential idea—is to send out high-frequency radio waves and detect distant objects by analyzing the reflections of these signals from the objects' surfaces. The principle of a laser printer is to use a computer-controlled laser to "write" images onto a copier drum. As with basic xerography, toner particles then stick to the parts of the drum that have been electrostatically written on, and are fused from there onto paper.

To be brought into physical reality a principle needs to be expressed in the form of physical components. In practice this means that a technology consists of a main assembly: an overall backbone of the device or method that executes its base principle. This backbone is supported by other assemblies to take care of its working, regulate its function, feed it with energy, and perform other subsidiary tasks. So the primary structure of a technology consists of a main assembly that carries out its base function plus a set of sub-assemblies that support this.

Let us look at the jet engine this way. The principle is simple enough: burn fuel in a constant flow of pressurized air and expel the resulting high-velocity gas backward. (By Newton's third law this produces an equal-and-opposite forward force.) To carry this out the engine uses a main assembly that consists of five main systems: intake, compressor, combustor, turbine, and exhaust nozzle. Air enters the intake and flows into the compressor assembly—essentially a series of large fans—which work to pump up its pressure. The now high-pressure airflow enters the combustor where it is mixed with fuel and ignited. The resulting high-temperature gas turns a set of turbines which drive the compressor, then expands through the nozzle section at high velocity to produce thrust. (Modern turbofan engines have a large fan in front, also driven by the turbines, that provides much of the thrust.)

These parts form the central assembly. Hung off this, and greatly complicating things, are many subsystems that support its main functions: a fuel supply system, compressor anti-stall system, turbine blade cooling system, engine instrument system, electrical system, and so on. All these assemblies and subsystems communicate with each other: the output of the fuel supply system becomes the input to the combustion system; the compressor signals its performance to the instrument system. To facilitate this the assemblies are connected by a complex labyrinth of pipes and electrical wires that act to hand off their functions for the use of other systems.

The setup is not different in a computer program. Again a base principle is used—the central concept or logic behind the program. This is implemented by a main set of instructional building blocks or functions—appropriately enough called "Main" in some computer languages. These call on other subfunctions or subroutines to support their workings. A program that sets up a graphic window on a computer display calls on subfunctions to create the window, set its size, set its position, display its title, fetch its content, bring it to the front of other windows, and delete it when it is done with. Such subfunctions call or make use of each other if certain conditions are met, so that the various subparts of a program are in constant interaction—constant conversation—with each other, just as they are in a jet engine. Physically, a jet engine and a computer program are very different things. One is a set of material parts, the other a set of logical instructions. But each has the same structure. Each is an arrangement of connected building blocks that consists of a central assembly that carries out a base principle, along with other assemblies or component systems that interact to support this.

Whether within a jet engine or a computer program, all parts must be carefully balanced. Each must be able to perform within the constraints—the range of temperatures, flow rates, loads, voltages, data types, protocols—set by the other parts it interacts with. And each in turn must set up a suitable working environment for the component assemblies that depend on it. In practice this means that difficult tradeoffs must be made. Each module or component must provide just the right power, or size, or strength, or weight, or performance, or data structure to fit with the rest. Each must therefore be designed to fit in a balanced way with the other parts.

Together these various modules and their connections form a working *architecture*. To understand a technology means to understand its principle, and how this translates into a working architecture.

Why Modularity?

We have now established a common structure for technologies. They consist of parts organized into component systems or modules. Some of these form a central assembly, others have supporting functions, and these themselves may have subassemblies and subparts. Of course there is no law that says that the components of a technology must be clustered into assemblies, into functional groupings, at all. We could easily imagine a technology that was put together completely from individual components. Yet beyond very simple technologies, I know of no such technologies.

Why should this be so? Why should technologies be organized from assemblies as well as from individual parts?

Several years ago, Herbert Simon told a classic parable of two watchmakers. Each assembles watches of 1,000 parts. The watchmaker named Tempus does this from individual pieces, but if he is interrupted or drops an unfinished watch, he must start again. Hora, by contrast, puts his watches together from 10 assemblies, each consisting of 10 subassemblies, each consisting of 10 parts. If he pauses or is interrupted, he loses only a small part of the work. Simon's point is that grouping parts into assemblies gives better protection against unexpected shocks, more ease of construction, and more ease of repair. We can extend this and say also that grouping parts allows for separate improvement of the component organs of a technology: these can be split out from the whole for specialized attention and modification. It allows for separate testing and separate analysis of working functions: their assemblies can be "slid out" to be probed or replaced separately without dismantling the rest of the technology. And it allows for swift reconfiguration of the technology to suit different purposes: different assemblies can be switched in and out when needed.

Separating technologies into functional groupings also simplifies the process of design. If designers were to work with tens of thousands of individual parts they would drown in a sea of details. But if instead they partition a technology into different building blocks—the arithmetic processing element of a computer, the memory system, the power system—they can hold these in their minds, concentrate on them separately, and see more easily how these larger pieces fit together to contribute to the working of the whole. Partitioning technologies into groupings or modules corresponds to "chunking" in cognitive psychology, the idea that we break down anything complicated (the Second World War, say) into higher-level parts or chunks (the lead-up to the war, the outbreak of war, the invasion of the Soviet Union, the war in the Pacific, etc.) that we can more easily understand and manipulate.

It costs something—mental effort at the very least—to partition the components of a technology into separate functional units. So it pays to divide a technology into such modules only if they are used repeatedly—only if there is sufficient volume of use. There is a parallel here with what Adam Smith said about the division of labor. Smith pointed out that it pays to divide a factory's work into specialized jobs, but only if there is sufficient volume of production. Modularity, we can say, is to a technological economy what the division of labor is to a manufacturing one; it increases as given technologies are used more, and therefore as the economy expands. Or, to express the same thing in good Smithian language, the partition of technologies increases with the extent of the market.

The way a functional unit is organized changes too as it becomes more used. A module or assembly begins typically as a loose grouping of individual parts that jointly execute some particular function. Later in its life, the grouping solidifies into a specially constructed unit. DNA amplification (a process that creates billions of copies from a small sample of DNA) in its earliest days was a loosely put together combination of laboratory techniques. These days it has become encapsulated within specially constructed machines. This is a general rule: what starts as a series of parts loosely strung together, if used heavily enough, congeals into a self-contained unit. The modules of technology over time become standardized units.

Recursiveness and Its Consequences

We are now seeing quite a lot of structure within technologies. But we are not quite finished yet. Structure has another aspect. From our combination principle, a technology consists of component parts: assemblies, systems, individual parts. We could therefore conceptually break a technology into its functional components (ignoring whether these are supporting or central) starting from the top down.

Doing this would give us the overall technology, its main assembles, subassemblies, sub-subassemblies, and so on, until we arrive at the elemental parts.

The hierarchy that forms this way is treelike: the overall technology is the trunk, the main assemblies the main branches, their subassemblies the sub-branches, and so on, with the elemental parts the furthest twigs. (Of course it is not a perfect tree: the branches and subbranches—assemblies and subassemblies—interact and cross-link at different levels.) The depth of this hierarchy is the number of branches from trunk to some representative twig. It would be two for Tempus's technology: the overall watch and elemental parts. For Hora it would be four: the watch, its main assemblies, subassemblies, and elemental parts. Real-world technologies can be anywhere from two to ten or more layers deep; the more complicated and modular the technology, the deeper the hierarchy.

So far this does not tell us anything very general about structure except that it is hierarchical. But we can say more. Each assembly or subassembly or part has a task to perform. If it did not it would not be there. Each therefore is a means to a purpose. Each therefore, by my earlier definition, is a technology. This means that the assemblies, subassemblies, and individual parts are all executables—are all technologies. It follows that a technology consists of building blocks that are technologies, which consist of further building blocks that are technologies, which consist of yet further building blocks that are technologies, with the pattern repeating all the way down to the fundamental level of elemental components. Technologies, in other words, have a recursive structure. They consist of technologies within technologies all the way down to the elemental parts.

Recursiveness will be the second principle we will be working with. It is not a very familiar concept outside mathematics, physics, and computer science, where it means that structures consist of components that are in some way similar to themselves. In our context of course it does not mean that a jet engine consists of systems and parts that are little jet engines. That would be absurd. It means simply that a jet engine (or more generally, any technology) consists of component building blocks that are also technologies, and these consist of sub-parts that are also technologies, in a repeating (or recurring) pattern.

Technologies, then, are built from a hierarchy of technologies, and this has implications for how we should think of them, as we will see shortly. It also means that whatever we can say in general about technologies-singular must hold also for assemblies or subsystems at lower levels as well. In particular, because a technology consists of main assembly and supporting assemblies, each assembly or subsystem must be organized this way too.

So far, recursiveness sounds abstract. But when we look at it in action, it becomes perfectly concrete. Consider a technology of some complication: the F-35 Lightning II aircraft. (Military examples are useful because they provide for deep hierarchies.) The F-35 is a fighter aircraft that comes in conventional, short-takeoff-and-vertical-landing, and carrier-based variants. The version I have in mind is the naval carrier one, the F-35C. It is single-engined, supersonic, and stealthy, meaning it possesses a very small radar signature that makes it hard to detect. The F-35C has the angled-surface appearance typical of stealth aircraft. Yet it is a sleek affair, with wings that are large relative to its fuselage and a tail section or empennage that sticks out behind in two sections, carrying stabilizers angled vertically in a V.

The reason for these particulars is that the F-35C needs to pull off a set of design objectives that conflict. It needs to be structurally strong and heavy enough to withstand the high forces of carrier launches and tailhook-arrested landings, yet preserve high maneuverability and long-range fuel performance. It needs to have excellent low-speed control for carrier landings, yet be able to fly at more than 1.6 times the speed of sound. And it needs to have the angled surfaces that make it almost undetectable to radar, yet fly properly.

The F-35C has multiple purposes: to provide close air support, intercept enemy aircraft, suppress enemy radar defenses, and take out ground targets. It is therefore a means, a technology—an executable.

What would we see if we followed the F35C's hierarchical tree outward to the same twig? We can break out the F-35C's main assemblies as: the wings and empennage; the powerplant (or engine); the avionics suite (or aircraft electronic systems); landing gear; flight control systems; hydraulic system; and so forth. If we single out the powerplant (in this case a Pratt & Whitney F135 turbofan) we can decompose it into the usual jet-engine subsystems: air inlet system, compressor system, combustion system, turbine system, nozzle system. If we follow the air inlet system it consists of two boxlike supersonic inlets mounted on either side of the fuselage, just ahead of the wings. The supersonic inlets adjust intake air velocity so that the engine experiences similar conditions at takeoff speed or at Mach 1.6. To do this normal supersonic inlets use an interior assembly of movable plates. The F-35C instead uses a cleverly designed protrusion in the fuselage just ahead of the inlets, called a DSI (diverterless-supersonic-inlet) bump, that preadjusts airflow and helps control shock waves. We could follow this further: the DSI-bump assembly consists of certain metal-alloy substructures. One of these . . .

But we have arrived at the bottom of the hierarchy, at the elemental level of executables. Notice, beneath the jargon of the individual assemblies, the recurring theme: systems that consists of systems, interconnected, interacting, and mutually balanced. Executables that consist of executables, or technologies that consist of technologies, hierarchically arranged all the way down to the level of individual elements.

We could follow the hierarchy of executables upward as well. The F-35C is an executable within a larger system, a carrier air wing. This consists of several fighter squadrons along with other support aircraft, whose purpose is to provide both striking power and electronic warfare capabilities. The air wing is part of a larger system; it deploys aboard a carrier. The carrier is an executable too: its purpose is (among others) to store, launch, and retrieve aircraft. It in turn typically will be an executable within a larger system, a carrier battle group. This, as its name suggests, is a combination of ships—guided-missile cruisers and frigates, destroyers, escort and supply ships, and nuclear submarines—grouped around the carrier. And it has varying purposes (missions in military jargon): showing presence, projecting force, escorting commercial vessels. Therefore the carrier group too is an executable. It in its turn may be part of a still larger system, a theater-of-war grouping: a carrier group supported by land-based air units, airborne refueling tankers, Naval Reconnaissance Office satellites, ground surveillance units, and marine aviation units. This larger level theater-of-war grouping may be fluid and changing, but it is also a means—it carries out naval assignments. So it too is an executable. It too is a technology if we care to look at it that way.

Thus the whole theater-of-war system I have described is a hierarchy of "technologies" in operation. It is nine or more layers deep. We can enter this hierarchy at any level—at any component system—and find it too is a hierarchy of executables in action. The system is self-similar. Of course executables at different levels differ in type and theme and appearance and purpose. The air inlet system is not a miniature version of the F-35C. But at any level each system and subsystem is a technology. Each is a means, and an executable. Given this, self-similarity carries up and down a recursive hierarchy that is many layers deep.

We can draw a number of general lessons from this example.

The standard view—I am talking about the one most technology thinkers have taken—sees a technology as something largely self-sufficient and fixed in structure, but subject to occasional innovations. But this is true of technologies only if we think of them in the abstract, isolated in the lab, so to speak. "In the wild"—meaning in the real world—a technology rarely is fixed. It constantly changes its architecture, adapts and reconfigures as purposes change and improvements occur. A carrier's jet fighters may act as more or less independent components one day. The next they may be assigned to the protection of a radar surveillance aircraft, becoming part of a new temporary grouping. New structures, new architectures, at any level can form quickly and easily, as needs require. In the real world, technologies are highly reconfigurable; they are fluid things, never static, never finished, never perfect.

We also tend to think of technology as existing at a certain scale in the economic world. If the technology is a traditional processing one (the basic oxygen process for producing steel), this is largely the scale of factories; if it is a device (a mobile phone), this is largely the scale of products. But in our example, the theater-of-war group as a whole is a technology, and so too is the smallest transistor within one of its aircraft control systems, and so are all the components in between. There is no characteristic scale for technology.

This does not mean there is no sense of "higher" and "lower" in a technology. Technologies at a higher level direct or "program" (as in a computer program) technologies at lower levels; they organize them to execute their purposes. The carrier "programs" its aircraft to execute its purposes. And the elements (technologies) at lower levels determine what the higher levels can accomplish. A theater-of-war grouping is limited by what its component carriers can achieve; and a carrier is limited by what its aircraft can achieve.

We can also see that every technology stands ready, at least potentially, to become a component in further technologies at a higher level. An F-35C in principle could stand on its own as a single aircraft with a single mission. But in operational practice it is very much part of a larger system, an air wing aboard a carrier, and it carries out its purposes within this larger context. This gives us backing for the assertion that all technologies stand by for use as potential components within other new technologies.

Recursiveness also carries a deeper implication. Very often in the world of technology, changes at one level must be accommodated by changes at a different level. The F-35C provides a different set of capabilities than its predecessor, the F/A-18 Hornet. And this means that the higher-level carrier systems that control and deploy it must also change the way they are organized. This theme of alterations in a technology requiring alterations at different levels is one we will see more of in the chapters ahead. It runs through all of technology.

I have been talking about the logical structure of technologies in this chapter, but I do not want to exaggerate what technologies have in common. A carrier group is a technology, but it is very different from, say, the process of distilling whisky. Still, granted their unique characters technologies do share a common anatomical organization. Each derives from a central principle and has a central assembly—an overall backbone of the device or method that executes this—plus other assemblies hung off this to make this workable and regulate its function. Each of these assemblies is itself a technology, and therefore itself has a central backbone and other subassemblies attached to this. The structure is recursive. The existence of this structure carries an important implication for us. Combination must work not just by bringing together a purpose with a concept or principle that matches it. It must provide a main set of assemblies or modules to execute this central idea. It must support this with further assemblies, and these again with further assemblies to support these. And all these parts and assemblies must be orchestrated to perform together harmoniously. Combination must necessarily be a highly disciplined process.

All of this will be useful later when we are looking at how technologies are brought into being and develop. But before we do so we still need to answer some of the deeper questions of the last chapter. What makes a technology a technology in the first place? What is technology in the deepest nature of its being? What, in other words, is the essence of technology? And what gives a particular technology its power? Certainly it cannot quite be its principle; that after all, is only an idea. It must be something else.

These questions, as we will see, can be answered by thinking about phenomena and how technologies make use of these. This is what we will do in the next chapter.

3

PHENOMENA

Archaeologists who want to establish the date of a particular site have a number of techniques they can use. If they find organic material, say the bones of an animal, they can use radiocarbon dating. If they find the remains of wooden structures, a post or lintel say, they can use dendrochronology, or tree-ring dating. If they find a firepit they can use archaeomagnetic dating.

Radiocarbon dating works because, when alive, an organism takes in carbon from the air or through the food chain; carbon contains small amounts of the radioactive isotope carbon-14, which decays into nonradioactive standard carbon at a constant rate; when the organism dies it ceases to ingest carbon, so the proportion of carbon-14 in its remains steadily decays. Measuring the relative amount of carbon-14 content therefore establishes a fairly accurate date for the specimen.

Dendrochronology works because tree rings vary in width season by season according to the rainfall received, and so trees that grow in a given climatic region and historical period show similar ring-width patterns. Comparing the ring pattern to a known and dated local ring pattern establishes exactly the years in which the wood in the structure was growing.

Archaeomagnetic dating works because the earth's magnetic field changes direction over time gradually in a known way. Clays or other materials in a firepit, when fired and cooled, retain a weak magnetism that aligns with the earth's field, and this establishes a rough date for the firepit's last use.

There are still other techniques: potassium-argon dating, thermoluminescence dating, hydration dating, fission-track dating. But what I want the reader to notice is that each of these relies on some particular set of natural effects.

That a technology relies on some effect is general. A technology is always based on some phenomenon or truism of nature that can be exploited and used to a purpose. I say "always" for the simple reason that a technology that exploited nothing could achieve nothing. This is the third of the three principles I am putting forward, and it is just as important to my argument as the other two, combination and recursiveness. This principle says that if you examine any technology you find always at its center some effect that it uses. Oil refining is based on the phenomenon that different components or fractions of vaporized crude oil condense at different temperatures. A lowly hammer depends on the phenomenon of transmission of momentum (in this case from a moving object—the hammer—to a stationary one—the nail).

Often the effect is obvious. But sometimes it is hard to see, particularly when we are very familiar with the technology. What phenomenon does a truck use? A truck does not seem to be based on any particular law of nature. Nevertheless it does use a phenomenon—or, I should say, two. A truck is in essence a platform that is self-powered and can be moved easily. Central to its self-powering is the phenomenon that certain chemical substances (diesel fuel, say) yield energy when burned; and central to its ease of motion is the "phenomenon" that objects that roll do so with extremely low friction compared with ones that slide (which is used of course in the wheels and bearings). This last "phenomenon" is hardly a law of nature; it is merely a usable—and humble—natural effect. Still it is a powerful one and is exploited everywhere wheels or rolling parts are used.

Phenomena are the indispensable source from which all technologies arise. All technologies, no matter how simple or sophisticated, are dressed-up versions of the use of some effect—or more usually, of several effects.

One way to convince yourself of this is to imagine you must measure something not easily measured. Suppose you are in space under zero gravity, and you must measure the mass of a small metal part. You cannot put it on a scale, or use it as a pendulum, or oscillate it on a spring—all these require gravity. You might think of oscillating it between two springs, or accelerating it somehow and measuring the force needed: you can indirectly measure mass this way. But notice what you are looking for is some phenomenon, some effect, that varies with the thing you are trying to measure. As with all technologies, you need some reliable effect to build your method from.

Examples such as this sound a bit like the physics puzzles traditionally used in undergraduate exams, the sort of question that requires you to find a building's height using a barometer, a ball of string, a pencil, and sealing wax. Here is an example that also has the flavor of a physics puzzle because it starts with a problem and works its way toward effects that can be used to solve it. It is a real-world technology that relies on four base effects.

The problem is how to detect planets that orbit distant stars (exoplanets, as they are called), and the technique I will describe was developed in the 1990s by the astronomers Geoffrey Marcy and Paul Butler. No existing telescope until recently could directly see exoplanets—they are too far away. So astronomers have been forced to seek indirect evidence of them. Marcy and Butler start from a simple but subtle phenomenon. Think of a star as floating out in space, many light-years distant. Planets provide a faint gravitational tug on the star they revolve around, and this causes the star to move back and forth in a regularly repeating way. The star's oscillation is slight, a few meters per second. And it is slow; it takes place over the planet's orbiting period of months and years. But if astronomers can detect it, they can infer the presence of a planet.

But how to detect this slow oscillation (or wobble, as astronomers call it)? Two further effects now enter. Light from a star can be split into a spectrum with its own distinctive bands of color (or light-frequencies); and these lines shift a little if the star is moving with respect to us (the famous Doppler effect). Putting these two effects together, Marcy and Butler can point a telescope at a chosen star, split its light into a spectrum and attempt to measure any wobble by looking for a back and forth shift in its spectral lines over a period of months or years. This sounds straightforward, but there is a difficulty. The shift this causes in the star's light spectrum is miniscule. If you think of a particular frequency line in the spectrum as corresponding to middle C on the musical scale, Marcy and Butler are looking for shifts corresponding to 1/100 millionth of the distance between C and C-sharp. How do you measure such infinitesimal shifts in a light spectrum?

Marcy and Butler—and this is their key contribution—use a fourth effect to do this. They arrange that the star's light be passed through an iodine-gas cell. Light passed through a gas cell is absorbed at particular frequencies characteristic of that gas, so the star's light spectrum shows black "absorption bands" not unlike a supermarket bar-code superimposed on it. The iodine cell is not moving, so its bands act as something like a fixed ruler against which it is possible to discern tiny shifts in a star's spectral lines as it moves closer to or away from the observer. The technique requires a great deal of refinement—it took Marcy and Butler nine years to perfect it. But it is sensitive enough to measure a star's motion down to the 10 meters or so per second required to infer the presence of exoplanets.

This technique has more the feel of a cobbled-together physics experiment—which is exactly what it is—than a highly elaborate commercial contrivance. But it illustrates the use of effects put together to achieve a purpose. Four effects in this case: that star-wobble is caused by the presence of a planet; that a star's light can be split into spectral lines; that these spectral lines shift if the star moves (wobbles) relative to us; and that light passed through a gas produces fixed absorption lines (that can act as a benchmark for any shift in a star's light spectrum). Captured and organized properly into a working technology, these effects have found about 150 new exoplanets.

I said earlier that a technology is based on a concept or principle. How does this fit with a technology's being based upon a phenomenon? Are a principle and a phenomenon the same thing? The answer is no—at least not as I am using the word "principle." A technology is built upon some principle, "some method of the thing," that constitutes the base of idea of its working; this principle in turn exploits some effect (or several) to do this. Principle and phenomenon are different. That certain objects—pendulums or quartz crystals—oscillate at a steady given frequency is a phenomenon. *Using* this phenomenon for time keeping constitutes a principle, and yields from this a clock. That high-frequency radio signals show a disturbance or echo in the presence of metal objects is a phenomenon. *Using* this phenomenon to detect aircraft by sending out signals and detecting their echoes constitutes a principle, and yields radar. Phenomena are simply natural effects, and as such they exist independently of humans and of technology. They have no "use" attached to them. A principle by contrast is the *idea of use of a phenomenon for some purpose* and it exists very much in the world of humans and of use.

In practice, before phenomena can be used for a technology, they must be harnessed and set up to work properly. Phenomena rarely can be used in raw form. They may have to be coaxed and tuned to operate satisfactorily, and they may work only in a narrow range of conditions. So the right combination of supporting means to set them up for the purpose intended must be found.

This is where the supporting technologies I spoke of in the previous chapter come in. Many of them, as I said, are there to provide for the base principle, feed it with energy, manage and regulate it. But many are also there to support the phenomena used and to arrange that they properly perform their needed function. Marcy and Butler's iodine cell must be held at exactly 50 degrees Celsius; a heating unit is needed to arrange this. The star's spectrum gets smeared out very slightly inside the spectrometer; computer methods are needed to correct for this. The earth itself is moving through space; further data-based algorithms are needed to correct for that. The star's light may show bursts of intensity; yet more computer methods are required to winnow out the true shift. All these needs require their own assemblies and components. The heating unit requires insulation and a temperature control assembly. The computer methods require specially written software. Getting the phenomena to work requires a plethora of assemblies and supporting technologies.

This gives us a deeper reason for technologies to consist of modules than the "chunking" for convenience of design or assembly I spoke of earlier. The subsystems used to render a phenomenon usable for a specific purpose are technologies, and they rely on their own phenomena. Therefore a practical technology consists of many phenomena working together. A radio receiver is not just a collection of parts. Nor is it merely a miniature signal processing factory. It is the orchestration of a collection of phenomena—induction, electron attraction and repulsion, electron emission, voltage drop across resistance, frequency resonance—all convened and set up to work in concert for a specific purpose.

Nearly always it is necessary to set up these phenomena and separate them by allocating them to separate modules. Within electronics you do not want to put an inductor close to a capacitor (that could cause unwanted oscillations). Within an aircraft powerplant you do not want to put combustion close to a device that measures, say, a pressure drop. So you allocate these phenomena to separate modules.

The Essence of Technology

We now have a more direct description of technology than saying it is a means to a purpose. A technology is a phenomenon captured and put to use. Or more accurately I should say it is a *collection* of phenomena captured and put to use. I use the word "captured" here, but many other words would do as well. I could say the phenomenon is harnessed, seized, secured, used, employed, taken advantage of, or exploited for some purpose. To my mind though, "captured and put to use" states what I mean the best.

This observation brings us back to the question I asked at the start of this book: What is the essence of technology? What, in its deepest nature, is technology? For me, the answer is what we have just arrived at: that a technology is a phenomenon captured and put to use. Or more usually, a set of phenomena captured and put to use. The reason this is central is that the base concept of a technology—what makes a technology work at all—is always the use of some core effect or effects. In its essence, a technology consists of certain phenomena programmed for some purpose. I use the word "programmed" here deliberately to signify that the phenomena that make a technology work are organized in a planned way; they are orchestrated for use.

This gives us another way to state the essence of technology. A technology is a programming of phenomena to our purposes.

This programming need not be obvious. And it need not be visible either, if we look at the technology from the outside. If we look at a jet engine from the outside all we can see is that it provides power. Considerable power. At this level the powerplant is just a means—a device that provides thrust—a remarkable device perhaps, but one we can still take for granted. If we look deeper into it, as say when its cowling is removed for maintenance, we can see it as a collection of parts—a tangle of pipes and systems and wiring and blades. The engine is now visibly a combination of executables. This is more impressive, but still something we can take for granted. But look deeper again, this time beneath what is visible. The powerplant is really a collection of phenomena "programmed" to work together, an orchestration of phenomena working together.

None of these phenomena is particularly esoteric; most are fairly basic physical effects. Let me list some of them (with the system that uses them in parenthesis): A fluid flow alters in velocity and pressure when energy is transferred to it (used in the compressor); certain carbon-based molecules release energy when combined with oxygen and raised to a high temperature (the combustion system); a greater temperature difference between source and sink produces greater thermal efficiency (again the combustion system); a thin film of certain molecules allows materials to slide past each other easily (the lubrication systems); a fluid flow impinging on a movable surface can produce "work" (the turbine); load forces deflect materials (certain pressure-measuring devices); load forces can be transmitted by physical structures (load bearings and structural components); increased fluid motion causes a drop in pressure (the Bernoulli effect, used in flow-measuring instruments); mass expelled at a velocity produces an equal and opposite reaction (the fan and exhaust propulsion systems). We could add many other mechanical phenomena. Then we could add the electrical and electronic phenomena, which are the basis of the engine's control, sensing, and instrumentation systems. And we could add still others, optical ones, for example. Each of these systems exploits some phenomenon, and because systems repeat fractally the overall device we are examining exploits phenomena all the way down to its elemental parts.

All these phenomena—scores of them—are captured, encapsulated in a myriad of devices, and replicated, some many thousands of times in as many thousands of identical components. That all these phenomena are caught and captured and schooled and put to work in parallel at exactly the right temperature and pressure and airflow conditions; that all these execute in concert with exactly the right timing; that all these persist despite extremes of vibration and heat and stress; that all these perform together to produce tens of thousands of pounds of thrust is not to be taken for granted. It is a wonder.

Seen this way, a technology in operation—in this case a jet engine—ceases to be a mere object at work. It becomes a metabolism. This is not a familiar way to look at any technology. But what I mean is that the technology becomes a complex of interactive processes—a complex of captured phenomena—supporting each other, using each other, "conversing" with each other, "calling" each other much as subroutines in computer programs call each other. The "calling" here need not be triggered in some sequence as in computing. It is ongoing and continuously interactive. Some assemblies are on, some are off; some operate continuously. Some operate in sequence; some operate in parallel. Some are brought in only in abnormal conditions.

For device technologies such as our aircraft powerplant, this calling is more likely to be continuous in time and parallel, as say when the combustion system "calls for," or executes under, a continuous supply of high-compression air for combustion. This is more like the concurrent operation of computer algorithms that "converse" with each other, that are always on, and interact continuously. For method technologies—industrial processes or algorithms, say—the calling is more apt to be in sequence. It is more like the operation of a standard sequential-operation computer program, but still an interactive dynamic process—a metabolism.

Viewed this way, a technology is more than a mere means. It is a programming of phenomena for a purpose. A technology is an orchestration of phenomena to our use.

There is a consequence to this. I said in Chapter 1 that technology has no neat genetics. This is true, but that does not mean that technology possesses nothing quite like genes. Phenomena, I propose, are the "genes" of technology. The parallel is not exact of course, but still, I find it helpful to think this way. We know that biology creates its structures—proteins, cells, hormones, and the like—by activating genes. In the human case there are about 21,000 of these, and the number does not vary all that much between fruit flies and humans, or humans and elephants. Individual genes do not correspond to particular structures; there is no single gene that creates the eye or even eye color. Instead, modern biology understands that genes collectively act as the elements of a programming language for the creation of a huge variety of shapes and forms. They operate much as the fixed set of musical tones and rhythms and phrases act as a programming language for the creation of very different musical structures. Organisms create themselves in many different shapes and species by using much the same set of genes "programmed" to activate in different sequences.

It is the same with technology. It creates its structures—individual technologies—by "programming" a fixed set of phenomena in many different ways. New phenomena—new technological "genes"—of course add to this fixed set as time progresses. And phenomena are not combined directly; first they are captured and expressed as technological elements which are then combined. There are probably fewer phenomena than biological genes in use, but still, the analogy applies. Biology programs genes into myriad structures, and technology programs phenomena to myriad uses.

Purposed Systems

This is a good place to deal with a problem the reader may already have thought of. I defined a technology as a means to a purpose. But there are many means to purposes and some of them do not feel like technologies at all. Business organizations, legal systems, monetary systems, and contracts are all means to purposes and so my definition makes them technologies. Moreover, their subparts—such as subdivisions and departments of organizations, say—are also means to purposes, so they seem to share the properties of technology. But they lack in some sense the "technology-ness" of technologies. So, what do we do about them? If we disallow them as technologies, then my definition of technologies as means to purposes is not valid. But if we accept them, where do we draw the line? Is a Mahler symphony a technology? It also is a means to fulfill a purpose—providing a sensual experience, say—and it too has component parts that in turn fulfill purposes. Is Mahler then an engineer? Is his second symphony—if you can pardon the pun—an orchestration of phenomena to a purpose?

When I came on this problem I was tempted to dismiss it and restrict technologies to means that are devices and methods. But I do not want to do this. If we can get comfortable with the idea that *all* means—monetary systems, contracts, symphonies, and legal codes, as well as physical methods and devices—are technologies, then the logic of technology would apply to a much wider class of things than the devices and methods I have been talking about, and that would greatly enlarge the scope of our argument.

So let us first look at why some technologies do not feel like standard technologies. Money is a means to the purpose of exchange, and therefore qualifies as a technology. (I am talking here about the monetary system, not the coins and paper notes that we carry.) Its principle is that any category of scarce objects can serve as a medium for exchange: gold, government-issued paper, or when these fail, cigarettes and nylons. The monetary system makes use of the "phenomenon" that we trust a medium has value as long as we believe that others trust it has value and we believe this trust will continue in the future. Notice the phenomenon here is a behavioral, not a physical one. This explains why money fulfills the requirements of a technology but does not feel like a technology. It is not based on a physical phenomenon. The same can be said for the other nontechnology-like technologies I listed above. If we examine them we find they too are based upon behavioral or organizational "effects," not on physical ones.

So we can say this: Conventional technologies, such as radar and electricity generation, feel like "technologies" because they are based upon physical phenomena. Nonconventional ones, such as contracts and legal systems, do not feel like technologies because they are based upon nonphysical "effects"—organizational or behavioral effects, or even logical or mathematical ones in the case of algorithms. The signature of standard technology—what makes us recognize something as a technology—is that it is based upon a physical effect.

So far so good. But this still does not tell us what we should do about these non-physically-based technologies. We can certainly

THE NATURE OF TECHNOLOGY

admit them as technologies for the argument of this book, and I will do so. But we can also recognize that in daily life we do not usually see them as technologies. A Mahler symphony is normally just an aesthetic experience, and a software company is normally just an organization. But we should remember that these too are "technologies" if we choose to see them this way. Mahler is very deliberately "programming" phenomena in our brains. To be specific he is arranging to set up responses in our cochlear nuclei, brain stems, cerebellums, and auditory cortices. At least in this sense Mahler is an engineer.

We can make all this a good deal more comfortable by adopting a simple stratagem. We can recognize that all along in the book we have really been talking about a class of systems: a class I will call *purposed systems*. This is the class of all means to purposes, whether physically or non-physically based. Some means—radar, the laser, MRI—we can prefer to think of as technologies in the traditional sense. Others—symphonies and organizations—we can prefer to think of as purposed systems, more like first cousins to technology, even if formally they qualify as technologies. That way we can talk about narrower physical technologies for most of our discussion, but we can also extend to the non-physical purposed-system ones when we want.

All this seems to be a digression. But it does establish the scope of what we are talking about. We can admit musical structures, money, legal codes, institutions, and organizations—indeed *all* means or purposed systems—to the argument even if they do not depend upon physical effects. With suitable changes, the logic I am laying out also applies to them.

Capturing Phenomena

I have been arguing in this chapter that phenomena are the source of all technologies and the essence of technology lies in orchestrating them to fulfill a purpose. This raises the obvious question: how do we uncover phenomena and capture them for use in the first place?

I like to think of phenomena as hidden underground, not available until discovered and mined into. Of course they are not scattered about randomly underground, they cluster into related families, each forming a seam or vein of useable effects—the optical phenomena, the chemical phenomena, the electrical phenomena, the quantum phenomena, and so on—and the members of these groups are unearthed piece by piece, slowly and haphazardly, over a lengthy period. Effects nearer the surface, say that wood rubbed together creates heat and thereby fire, are stumbled upon by accident or casual exploration and are harnessed in the earliest times. Deeper underground are seams such as the chemical effects. Some of these are uncovered by early adepts, but their full uncovering requires systematic investigation. The discovery of the most deeply hidden phenomena, such as the quantum effects of nuclear magnetic resonance or tunneling or stimulated emission, require a cumulation of knowledge and modern techniques to reveal themselves. They require modern methods of discovery and recovery—they require, in other words, modern science.

How then does science uncover novel effects? Certainly it cannot set out to uncover a new effect directly; that would be impossible—a new effect is by definition unknown. Science uncovers effects by picking up intimations of something that operates differently than expected. An investigator notices something here, a lab ignores something there but picks up the trace later. Röntgen stumbled on X-rays by observing that when he operated a Crookes tube (essentially a cathode-ray tube), a cardboard screen coated with barium platinocyanide several feet away subtly glowed. Sometimes the spoor is picked up by theoretical reasoning. Planck "discovered" the quantum (or, more precisely, introduced the notion that energy is quantized) via a theoretical attempt to explain the spectrum of blackbody radiation. A new effect declares itself always as the byproduct of some endeavor. The phenomenon of complementary base pairing in DNA (where the chemical base A is matched with T, and C with G), which is central to all genetic technologies, was uncovered by Watson and Crick as a byproduct of their attempt to model DNA's physical structure.

This makes it look as if science uncovers new effects piece by piece, each independently. Actually, this is not quite the case. Earlier effects discovered within a particular family of phenomena lead to the discovery of later ones. In 1831 Faraday discovered electromagnetic induction. He took an iron ring and wound copper wire around one side of it, and this set up a powerful magnetic field in the ring when he connected a battery to the coil. He had wound a second coil of wire around the ring and noticed that when he switched the magnetic field on, this caused or "induced" a current in the second coil that he could detect using a magnetic compass: "Immediately, a sensible effect on needle." The needle also moved when he disconnected the battery. Faraday had discovered that a changing magnetic field induced an electrical current in a nearby conductor.

We could say Faraday used scientific insight and experiment to do this, and certainly that would be true. But if we look more closely, we see that he uncovered induction using instruments and understandings that had grown from effects uncovered earlier. The battery used electrochemical effects discovered earlier. The effect that coils produce stronger magnetic fields when wound around a magnetic material like iron had recently been discovered by Gerrit Moll of Utrecht. And his current-detecting device exploited Ørsted's discovery of eleven years earlier that an electrical current deflects a magnetic needle. Faraday's discovery was possible because earlier effects had been harnessed into existing methods and understandings.

This is general with phenomena. As a family of phenomena is mined into, effects uncovered earlier begin to create methods and understandings that help uncover later effects. One effect leads to another, then to another, until eventually a whole vein of related phenomena has been mined into. A family of effects forms a set of chambers connected by seams and passageways, one leading to another. And that is not all. The chambers in one place—one family of effects—lead through passageways to chambers elsewhere—to different families. Quantum phenomena could not have been uncovered without the prior uncovering of the electrical phenomena. Phenomena form a connected system of excavated chambers and passageways. The whole system underground is connected.

Once phenomena are mined, how are they translated into technologies? This is something we will look at in detail later. For now, let me just say that by their very nature phenomena *do* something, and they start to become harnessed when someone notices a potential use for what they do—it is not a huge step to see how the phenomenon of a changing magnetic field inducing an electric current could be translated into a means for generating electricity.

Not every phenomenon of course is harnessable for use, but when a family of phenomena is uncovered, a train of technologies follows. The main electrical phenomena were uncovered between 1750 and 1875—electrostatic effects, galvanic action, the deflection of charges by electric and magnetic fields, induction, electromagnetic radiation, and glow discharges. With the capturing and harnessing of these for use, a parade of methods and devices followed—the electric battery, capacitors and inductors, transformers, telegraphy, the electrical generator and motor, the telephone, wireless telegraphy, the cathode ray tube, the vacuum tube.

This buildout happened slowly as earlier effects in the form of instruments and devices helped uncover later ones. In this way the uncovering of phenomena builds itself out of itself. Phenomena cumulate by bootstrapping their way forward: effects are captured and devices using these effects are built and uncover further effects.

Technology and Science

Of course all this requires understanding, or at the very least knowledge of effects and how to work with them. Where then does this knowledge fit in? It is very obvious that knowledge—information, facts, truths, general principles—is necessary for all that we have been looking at. So let us ask a sharper version of the same question. Where does the formal knowledge of phenomena—science—fit in? How is science connected to the use of phenomena? For that matter how is science connected to technology itself?

Science is necessary for the unearthing of modern phenomena, the more deeply hidden clusters of effects, and for forming technologies from these. It provides the means for observing effects; the understandings needed for working with them; the theories predicting how they will behave; and often the methods to capture them for use. So it is necessary for all our dealings with modern phenomena. This is fine—unobjectionable even. But this leads us straightaway into contested territory. Science appears to uncover novel effects while technology exploits these, so it seems that science discovers and technology applies.

Does this mean that technology is applied science? The idea certainly has its supporters. "Technology," declared the late distinguished engineering professor John G. Truxal, "is simply the application of scientific knowledge to achieve a specified human purpose."

Is technology indeed the application of science? *Simply* that? I believe it is not, or at least that the truth is a good deal more complicated than this. For one thing, in the past many technologies—powered flight is one—came into being with little science at all. Indeed, it is only since the mid-1800s that technology has been borrowing on a large scale from science at all. The reason science arrived in technology about this time was not just that it provided more insight and better prediction of results. It was that the families of phenomena beginning to be exploited then—the electrical and chemical ones, for example—acted on a scale or in a world not accessible to direct human observation without the methods and instruments of science. To set up a means to send messages by the large wooden two-armed signaling devices of Napoleon's day merely required common sense; to set up a means to send them by electricity (in some form of telegraph) required systematic knowledge of electrical phenomena. Technologists use science because that is the only way we understand how phenomena at the deeper layers work.

But this still does not mean that technologists reach into scientific ideas and simply apply them. Technologists use scientific ideas much as politicians use the ideas of defunct political philosophers. They exploit them day-to-day without much awareness of the details of their origin. This is not because of ignorance. It is because ideas that originate in science become digested over time into the bodies of technology themselves—into electronics or biotechnology. They mingle in these fields with experience and special applications to create their own indigenous theories and practice. So it is naïve to say that technology "applies" science. It is better to say it builds both from science and from its own experience. These two cumulate together, and as this happens science organically becomes part of technology. Science is deeply woven into technology.

But equally deeply, technology is woven into science. Science achieves its insights by observation and reasoning, but what makes these possible is its use of methods and devices. Science is in no small part the probing of nature via instruments and methods—via technology. The telescope created the modern science of astronomy every bit as much as the reasonings of Copernicus and Newton did. And Watson and Crick could not have uncovered the structure of DNA (and therefore the phenomenon of complementary base pairing) without the methods and devices of X-ray diffraction and the biochemical methods necessary for the extraction and purification of DNA. Without instruments for the observing and understanding of phenomena—without the microscope, methods of chemical assaying, spectroscopy, the cloud chamber, instruments for measuring magnetic and electrical effects, X-ray diffraction methods, and a host of other modalities—modern science would not exist at all. All these are technologies although we do not normally think of them this way. And science builds its understandings from these.

What about scientific experiments? Are these related to technology too? Certainly some are mere prospectings in the hope of lucky discoveries. But the serious ones are systematic probes into the workings of nature and are undertaken always with a definite purpose in mind. They are therefore means to human purposes; they are method technologies, encased or embodied in physical apparatus. When Robert Millikan conducted his famous oil-drop experiments in 1910 and 1911, his purpose was to measure the electric charge on an electron. He constructed—put together—a method to do this. The idea, the base concept, was to attach a very small number of electrons to a tiny droplet of oil. Using an electric field of known strength, he could control the motion of the charged droplet (charged particles are attracted or repelled in an electric field), allowing it to float suspended against gravity or to move slowly up or down between two crosshairs. By measuring the motion of the drop in the known electric field (and separately figuring its size), Millikan could compute the charge on the electron.

It took Millikan five years to perfect his scheme. He put his method through several versions. In the ones he based his definitive measurements on, he used an atomizer to produce the microscopic oil droplets and arranged that negative electrical charges (electrons) attached themselves to these. He could then look through a horizontal microscope, pick out a particular droplet and allow it to fall under gravity (and no electrical field) between the two horizontal hairlines. The timing of the droplet's movement through air, using Stokes's equation for motion in a fluid, would later establish its size. He could then switch on the positive electrical field. This would cause the negatively charged droplet either to drift upward or stay still, balanced by gravity. From the droplet's speed of motion (or the voltage necessary to keep it still) and from knowing its size, he could calculate its charge. Observing dozens of droplets this way, Millikan found their charge varied in integral amounts, corresponding to the number of electrons that had attached themselves to them. The smallest integral value was the charge on a single electron.

Like all beautiful experiments, Millikan's had an elegance and simplicity. But notice what it really was. It was a means to a purpose:

a technique or method technology (notice the stages in it). These stages took place within a construction of component parts—atomizers, chambers, charged plates, batteries, ion source, observing microscope—which were themselves means to purposes. In fact, what is striking about Millikan's work is his attention to construction. His struggle throughout the five years to refine the measurement was a technological one: he needed to constantly refine the parts—the assemblies—of his apparatus piece by piece. Millikan was investigating a phenomenon and he was doing this in a perfectly scientific way. But he had constructed a method to do this: a technology.

Science probes in exactly this way. It uses technologies in the form of instruments and experiments to answer specific questions. But of course it builds itself upon more than these. It builds its understanding from its scientific explanations, from its reasonings and theories about how the world works. Surely, we can say, these at least are far away from technology.

Well, not quite. Explanations certainly do not feel like technologies. But they *are* constructions with a purpose. Their purpose is to make clear the workings of some observed feature of the world, and their components are conceptual elements that combine according to accepted rules. All explanations are constructions from simpler parts. When Newton "explains" planetary orbits, he constructs a conceptual version of them from simpler parts, from the elemental ideas of mass and of gravity acting between masses, and from his laws of motion. He constructs, if you like, a mathematical "story" about how planets orbit, from the accepted components and rules. It would be stretching things to call Newton's explanation, or other explanatory theories for that matter, technologies. We can more comfortably see them, again, as purposed systems—forms of, or cousins to, technology.

From all this it follows that science not only uses technology, it builds itself from technology. Not the standard set of technologies such as bridges and steel production methods and shipping, of course. Science builds itself from the instruments, methods, experiments, and conceptual constructions it uses. This should not be surprising. Science, after all, is a method: a method for understanding, for probing, for explaining. A method composed of many submethods. Stripped to its core structure, science is a form of technology.

This last thought may disturb—may horrify—my scientific readers. So let me be clear on what I am saying and not saying. I am saying that science forms from technologies: from instruments, methods, experiments, and explanations. These are the sinews of science. I am not saying—emphatically not—that science is the same as technology. Science is a thing of beauty in itself. Of surpassing beauty. And it is much more than its core structure of instruments, experiments, and explanations. It is a set of moral ideas: that nature is intrinsically knowable, that it can be probed, that causes can be singled out, that understandings can be gained if phenomena and their implications are explored in highly controlled ways. It is a set of practices and a way of thinking that includes theorizing, imagining, and conjecturing. It is a set of knowings—of understandings accumulated from observation and thought in the past. It is a culture: of beliefs and practices, friendships and exchanged ideas, opinions and convictions, competition and mutual help.

None of these can be reduced to standard technology. Indeed, it is possible to imagine science without technology—without the telescope, microscope, computer, measuring instruments—science based on thought and conjecture alone. But this very thought confirms my point. Science without technology would be a weak science. It would be little more than the thought-based science of the Greeks.

Where does all this leave us? We can say this. Technology builds from harnessing phenomena largely uncovered by science. And equally science builds from technology—or, better to say, forms from its technologies—from the use of the instruments and methods and experiments it develops. Science and technology co-evolve in a symbiotic relationship. Each takes part in the continued creation of the other, and as it does, takes in, digests, and uses the other; and in so doing becomes thoroughly intermingled with the other. The two cannot be separated, they rely completely on one another. Science is necessary to uncover and understand deeply buried phenomena, and technology is necessary to advance science.

Recently the economic historian Joel Mokyr has pointed out that technologies issue forth as human knowledge is gained. Knowledge painstakingly built up over the last four hundred years, he says, together with the social and scientific institutions that promoted and helped diffuse this knowledge, have laid the foundations for the industrial revolution and for modern technology.

I believe this. In fact, I believe along with Mokyr that it is almost impossible to exaggerate the importance of knowledge. You cannot properly design a modern jet engine without knowledge of fluid flow over airfoils (in this case the turbine and compressor blades). But I would express Mokyr's observation differently. As phenomena are uncovered and explored, they begin to be surrounded by a penumbra of understandings about their workings. These understandings—this theory and knowledge—help greatly in developing technologies from them. In fact nowadays they are indispensable. This is because phenomena used in technology now work at a scale and a range that casual observation and common sense have no access to. Where once common sense could produce new devices for textile weaving, only detailed, systematic, codified, theoretical knowledge can produce new techniques in genetic engineering or in microwave transmission. Or in searching for exoplanets, for that matter. But that is only one half of the story. New instruments and methods form from these uncovered and understood phenomena. These technologies help build yet further knowledge and understanding and help uncover yet further phenomena. Knowledge and technology in this way cumulate together.

I have been talking about phenomena in this chapter and how technologies arise from them. What does this say about the buildout of technology over time?

Nature possesses many sets of phenomena and over centuries we have mined into these for our purposes: fire and metal working in the prehistoric past, optics in the seventeenth century, chemistry and electricity in the eighteenth and nineteenth centuries, quantum phenomena in the twentieth, and genetic effects in the late twentieth. Some of these effects have been captured and set to use in the form of technologies. These in turn become potential building blocks for the creation of yet further technologies. And some technologies (in the form of scientific instruments and methods) help uncover yet further phenomena. There is a nice circle of causality here. We can say that novel phenomena provide new technologies that uncover novel phenomena; or that novel technologies uncover new phenomena that lead to further technologies. Either way, the collectives of technology and of known phenomena advance in tandem.

None of this should be taken to mean that technologies always proceed directly from phenomena. Most technologies are created from building-block components that are several steps removed from any direct harnessing of an effect. The Mars Rover is put together from drive motors, digital circuits, communication systems, steering servos, cameras, and wheels without any direct awareness of the effects that lie behind these. The same is true for most technologies. Still, it is good to remember that all technologies, even planetary vehicles, derive ultimately from phenomena. All ultimately are orchestrations of phenomena.

One last thing. It should be clear by now that technologies cannot exist without phenomena. But the reverse is not true. Phenomena purely in themselves have nothing to do with technology. They simply exist in our world (the physical ones at least) and we have no control whatever over their form and existence. All we can do is use them where usable. Had our species been born into a universe with different phenomena we would have developed different technologies. And had we uncovered phenomena over historical times in a different sequence, we would have developed different technologies. Were we to discover some part of the universe where our normal phenomena did not operate, then our technologies would not operate, and we could only probe this part of the universe by literally inventing our way into it. This sounds like a science fiction scenario, but it takes place on a small scale closer to earth when we try to operate in space with only one effect missing: gravity. Methods for doing even the simplest things—drinking water, for instance—have to be rethought.

4

DOMAINS,

OR WORLDS ENTERED FOR WHAT

CAN BE ACCOMPLISHED THERE

As families of phenomena—the chemical ones, electrical ones, quantum ones—are mined into and harnessed, they give rise to groupings of technologies that work naturally together. The devices and methods that work with electrons and their effects—capacitors, inductors, transistors, operational amplifiers—group naturally into electronics; they work with the medium of electrons, and can therefore "talk" to each other easily. Components that deal with light and its transmission—lasers, fiber optic cables, optical amplifiers, optical switches—group into photonics; they pass photons, units of light energy, to each other for different operations. Such groupings form the technologies-plural I talked of in Chapter 2, and I want to talk more about them now. In particular, I will argue that each grouping forms a language within which particular technologies—particular devices and methods—are put together as expressions within that language.

Why do individual technologies cluster into groupings? Sharing a family of effects is one reason, but it is not the only one. Elements that share a common purpose group together: the cables of cable-stayed bridges require anchoring devices, and these in turn require certain heavy-duty bolts; so cables, anchors, and bolt assemblies go naturally together. Elements that share common physical-strength and scale characteristics cluster: beams and trusses and columns and steel girders and concrete slabs match in strength and scale and use; therefore they form the building blocks of structural engineering. And elements that repeatedly form subparts of useful combinations cluster. In genetic engineering, methods such as DNA purification, DNA amplification, radiolabeling, sequencing, cutting via restriction enzymes, cloning fragments, and screening for expressed genes, form a natural toolbox of building blocks from which particular procedures are put together.

Sometimes elements cluster because they share a common theory. The components of statistical packages—operations that summarize and analyze data, and perform statistical tests—often assume normally distributed sample populations and therefore work together under shared assumptions about the nature of the data they manipulate. What delineates a cluster of technologies is always some form of commonality, some shared and natural ability of components to work together.

I will call such clusters—such bodies of technology—*domains*. A domain will be any cluster of components drawn from in order to form devices or methods, along with its collection of practices and knowledge, its rules of combination, and its associated way of thinking.

It will be crucial for us to keep the distinction between individual technologies and domains clear if we want to be unambiguous later about how technologies come into being and evolve. Sometimes the difference looks blurry: radar (the individual system) and radar technology (the engineering practice) sound very much the same. But they are not.

A technology (individual, that is) does a job; it achieves a purpose—often a very particular purpose. A domain (technology-plural) does no job; it merely exists as a toolbox of useful components to be drawn from, a set of practices to be used. A technology defines a product, or a process. A domain defines no product; it forms a constellation of technologies—a mutually supporting set—and when these are represented by the firms that produce them, it defines an industry. A technology is invented; it is put together by someone. A domain—think of radio engineering as a whole—is not invented; it emerges piece by piece from its individual parts. A technology—an individual computer, say—gives a certain potency to whoever possesses it. A domain—the digital technologies—gives potential to a whole economy that can in time become transmuted into future wealth and political power.

And there is another difference. Where technology possesses a hierarchy of assemblies, subassemblies, and components; domains possesses subdomains, and within these still further sub-subdomains. Within electronics lie the subdomains of analog electronics and digital electronics, and within these lies the sub-subdomain of solid-state semiconductor devices; within this again lie the sub-sub-subdomains of gallium arsenide and of silicon devices.

One way to state the difference is to say that an individual technology is to a body of technology as a computer program is to a programming language.

Domaining

Design in engineering begins by choosing a domain, that is, by choosing a suitable group of components to construct a device from. When architects design a new office building, they may choose, for visual and structural reasons, glass-and-steel components over granite-and-masonry ones. That is, they may choose between different characteristic palettes of components, and we would want to call these domains too. Often this choice—I will call it *domaining* the device—is automatic. A marine radar designer will domain a master oscillator unthinkingly within the group of components we call electronics, simply because no other domain would be appropriate.

But often the choice requires thought. A designer who puts together a computer operating system these days will need to consider domaining it within the Linux versus Windows collections of functionalities. Of course, any technology of significant size is put together from several domains. An electric power station is constructed from component systems drawn from the building-construction, hydraulic, heavy electrical, and electronic domains.

In the arts, choice of domain is largely a matter of style: a composer will allow a theme to enter and leave different domains of the orchestra, the horn section or the string section, to achieve certain feelings and contrasts. Within technology, choice of domain decides not mood nor feeling, but rather the convenience and effectiveness of an assembly, what it can accomplish, how easily it can link with other assemblies, and what it will cost. Domaining in technology is a practical business.

Over time the choice of domain for a given purpose changes. Before the arrival of digitization, aircraft designers domained the systems that controlled wing and stabilizer surfaces within mechanical and hydraulic technologies. They used push rods, pull rods, cables, pulleys, and other mechanical linkages to connect the movements of the pilot's stick and rudder with hydraulic machinery that moved the aircraft's control surfaces. Then in the 1970s they began to redomain aircraft control digitally in a new technology-plural called fly-by-wire. The new systems sensed the pilot's actions and the aircraft's current motions and sent these as signals by electrical wire to a digital computer for processing. The computer then transmitted the needed adjustments, again by wire, to fast hydraulic actuators that moved the control surfaces.

Fly-by-wire allowed aircraft control systems that were much lighter and more reliable than the heavy mechanical ones that preceded them. They were also faster: the new controls could respond to changes within small fractions of a second. And they were "intelligent." Fly-by-wire could use computers to set controls more accurately than a human could and even override undesirable pilot decisions.

The new domain in fact made possible a new generation of military aircraft designed to be inherently unstable (in technology jargon, to have "relaxed static stability"). This gave an advantage. Just as you can maneuver an unstable bicycle more easily than a stable tricycle, you can maneuver an inherently unstable aircraft more easily than a stable one. The new controls could act to stabilize the aircraft much as a cyclist stabilizes a bicycle by offsetting its instabilities with countering motions. A human pilot could not react quickly enough to do this; under the earlier manual technology, an inherently unstable aircraft would be unflyable.

We could say in this case that aircraft controls underwent an "innovation." And indeed they did. But more accurately we should say that aircraft controls were domained differently—*redomained*. This distinction is important. Innovations in history may often be improvements in a given technology—a better way to architect domes, a more efficient steam engine. But the significant ones are new domainings. They are the expressing of a given purpose in a different set of components, as when the provision of power changed from being expressed in waterwheel technology to being expressed in steam technology.

The reason such redomainings are powerful is not just that they provide a wholly new and more efficient way to carry out a purpose. They allow new possibilities. In the 1930s, aircraft approaching England across the channel could be detected by large acoustic mirrors made from concrete, fifteen feet or more high. These focused sound from as far away as twenty miles and were manned by listeners with extremely sensitive hearing. By the outbreak of the second world war the same purpose was carried out by radar, which was effective at a much greater range. We could say that aircraft detection adopted (or indeed initiated) radar. But I prefer to say it took advantage of a new and powerful domain, radio engineering, that was sweeping into the world and whose components were finding many uses.

A change in domain is the main way in which technology progresses.

When a novel domain appears it may have little direct importance. Early radio was limited mainly to wireless telegraphy—sending messages without telegraph wires, often from ship to shore. Nevertheless, the appearance of a new domain constitutes not what can be done but, more than this, the potential of what can be done. And it stimulates the minds of the technologists of the day. In 1821 Charles Babbage and his friend, the astronomer John Herschel, were preparing mathematical tables for the Astronomical Society. The two men were comparing two previous versions of the same mathematical tables, each hand-calculated separately, and Babbage became exasperated at the high frequency of errors they were finding. "I wish to God," he said, "these calculations had been executed by steam." Babbage's declaration sounds quaint now, and in the end he designed his calculating devices not to be driven by steam but by hand-cranked gears and levers. But notice his appeal was not to a new device, but to a new domain. To the wonder-domain of his day. Steam in the 1820s defined a new world of the possible.

The domains of any period in fact define not just what is possible but that period's style. Think of the illustrations of spaceship travel you find in late Victorian science fantasy novels. I have in mind here a particular book, Jules Verne's *From the Earth to the Moon* with its original 1860s French illustrations. If you see a spacecraft and its launching device illustrated in this book, you recognize these immediately as belonging to Jules Verne's time not because of their shape or design. These are particular. You recognize them as mid-1800s because of the component sets they draw from: the iron-plate cladding of the craft; the artillery cannon that hurls it into space; the brick-and-wrought-iron structures that house the venture. Such component sets and the way they are used do not just reflect the style of the times, they *define* the style of the times.

An era does not just create technology. Technology creates the era. And so the history of technology is not just the chronicle of individual discoveries and individual technologies: the printing press, the steam engine, the Bessemer process, radio, the computer. It is also the chronicle of epochs—whole periods—that are defined by how their purposes are put together.

You can see this—or I should say, *feel* this—in a vivid way if you step into a well-designed museum display. Very often museums have a special room, or several, devoted to recapturing a particular moment of history. The Los Alamos Historical Museum has such a display, showing the artifacts and the technologies of the war years. You see chemical retorts, slide rules, Manhattan Project ID cards, a dressed-up dummy in a military police uniform, gasoline ration cards, scenes of old trucks and jeeps. All these have to do with the characteristic means—technologies—used to carry through the tasks of the day. And all these, more than anything else, define the times they are describing. The Los Alamos exhibit itself is modest, only two or three rooms, yet to step into it is to step into those times.

Domains do not just define the times, they also define the reach of the times. Imagine trying to draw a geological map from seismic data using the domains of Babbage's day. It would require the labors of a Babbage himself to design the apparatus, to put together a seismic analyzer that would render a map from the explosions' echoes. The resulting machine would be a wonder—with a listening horn perhaps, and attendant gunmetal gears and levers and dials and inked drawing pens. It would be slow. And complicated. And it would be specific to seismic exploration. The domains of Babbage's day—machines, railways, early industrial chemistry—had a short reach; they allowed a narrow set of possibilities. The domains of our day present wider possibilities. In fact, half of the effectiveness of a domain lies in its reach—the possibilities it opens up; the other half lies in using similar combinations again and again for different purposes. These act something like the pre-formed blocks of commonly used expressions kept at hand for ready use by old-fashioned typesetters (the French printers of the 1700s called them clichés), except that they are conceptual and not necessarily premade.

Design as Expression Within a Language

A new device or method is put together from the available components—the available vocabulary—of a domain. In this sense a domain forms a language; and a new technological artifact constructed from components of the domain is an utterance in the domain's language. This makes technology as a whole a collection of several languages, because each new artifact may draw from several domains. And it means that the key activity in technology—engineering design—is a form of composition. It is expression within a language (or several).

This is not a familiar way to look at design. But consider. There are articulate and inarticulate utterances in a language; the same goes for design. There are appropriate and inappropriate choices in language; the same goes for design. There is conciseness in language; the same goes for design. There are degrees of complication in what is expressed in language; the same goes for design. An idea expressed in language can be simple and expressed in a single sentence; or it may take up an entire book, with a central theme and subthemes supporting this. The same goes for design. For any willed purpose of expression in language there are many choices of utterance; similarly in technology for any purpose there is a wide choice of composition. And just as utterances in a language must be put together according to the rules of that language, so must designs be architected according to the rules of allowable combination in a domain.

I will call such rules a grammar. Think of a grammar as the "chemistry," so to speak, of a domain—the principles that determine its allowable combinations. I am using the word "grammar" here in one of its common meanings. Henry James spoke of "the grammar of painting," and the biochemist Erwin Chargaff says of his discoveries in DNA chemistry in 1949: "I saw before me in dark contours the beginning of a grammar of biology." James and Chargaff do not mean the properties of painting or molecular biology, but rather how painting's or biology's elements interrelate, interact, and combine to generate structures.

A domain's grammar determines how its elements fit together and the conditions under which they fit together. It determines what "works." In this sense there are grammars of electronics, of hydraulics, and of genetic engineering. And corresponding to the finer domains of these are subgrammars and sub-subgrammars.

Where do such grammars arise from? Well, of course, ultimately from nature. Behind the grammar of electronics lies the physics of electron motions and the laws of electrical phenomena. Behind the grammar of DNA manipulation lie the innate properties of nucleotides and of the enzymes that work with DNA. Grammars in large part reflect our understanding of how nature works in a particular domain. But this understanding comes not just from theory. It comes also from practical prescriptions that accumulate with experience: working temperature and pressure parameters, the settings for machines and instruments used, process timings, material strengths, clearance distances for assemblies—the thousand-and-one small pieces of knowledge that comprise the "cookery" of the art.

Sometimes this sort of knowledge is reducible to rules of thumb. The aircraft design community knows from years of experience that "for successful jet airplanes the ratio of thrust of the engines to weight of the loaded aircraft always comes out somewhere between about 0.2 and 0.3." But often knowledge cannot be distilled into such rules. Technology does not differ from the arts in this regard. Just as Cordon Bleu cooking in Paris cannot be reduced to a set of written principles, neither can expert electronic design. Grammars in both cooking and engineering exist not just as rules but as a set of unspoken practices taken for granted, a practical knowing of how to do things that may not be fully expressible in words. Grammars comprise a culture of experience, a craft of use. And as such they exist more in the minds of practitioners and in their shared culture than in textbooks. They may start as rules, but they end as a way of conceptualizing technologies, a way of thinking.

In fact, just as articulate expression within a spoken language depends on more than mere grammar (it depends upon deep knowledge of the words in the language and their cultural associations), so too articulate expression in technology depends on more than grammar alone. Articulate utterance in technology requires deep knowledge of the domain in question: a fluency in the vocabulary of components used; a familiarity with standard modules, previous designs, standard materials, fastening technologies; a "knowingness" of what is natural and accepted in the culture of that domain. Intuitive knowledge, cross communication, feeling, past use, imagination, taste—all these count.

Technology expert James Newcomb, talking of the business of providing energy efficiency services, says: "Doing it well entails knowledge of literally thousands of individual technologies, together with the capability to assimilate and optimally combine these technologies in particular applications, taking into consideration interactive effects, control systems, process implications, and the differential economics of energy and demand savings. It's the skill of a master chef, not a grocer's buyer."

Good design in fact is like good poetry. Not in any sense of sublimity, but in the sheer rightness of choice from the many possible for each part. Each part must fit tightly, must work accurately, must conform to the interaction of the rest. The beauty in good design is that of appropriateness, of least effort for what is achieved. It derives from a feeling that all that is in place is properly in place, that not a piece can be rearranged, that nothing is to excess. Beauty in technology does not quite require originality. In technology both form and phrases are heavily borrowed from other utterances, so in this sense we could say that, ironically, design works by combining and manipulating clichés. Still, a beautiful design always contains some unexpected combination that shocks us with its appropriateness.

In technology, as in writing or speech—or haute cuisine—there are varying degrees of fluency, of articulateness, of self-expression. A beginning practitioner in architecture, like a beginner at a foreign language, will use the same base combinations—the same phrases— over and over, even if not quite appropriate. A practiced architect, steeped in the art of the domain, will have discarded any notion of the grammar as pure rules, and will use instead an intuitive knowledge of what fits together. And a true master will push the envelope, will write poetry in the domain, will leave his or her "signature" in the habit-combinations used.

Mastery in a technology in fact is difficult to achieve because a technology grammar, unlike a linguistic one, changes rapidly. Technology grammars are primitive and dimly perceived at first; they deepen as the base knowledge that comprises them grows; and they evolve as new combinations that work well are discovered and as the daily use of working designs reveals difficulties. There is never closure to them. As a result, even adepts can never fully keep up with all the principles of combination in their domain.

One result of this heavy investment in a domain is that a designer rarely puts a technology together from considerations of all domains available. The artist adapts himself, Paul Klee said, to the contents of his paintbox. "The painter . . . does not fit the paints to the world. He fits himself to the paint." As in art, so in technology. Designers construct from the domains they know.

Worlds Entered

A domain or body of technology, as I said, provides a language for expression, a vocabulary of components and practices designers can draw from. Computation (or digital technology) is a collection—an extremely large vocabulary—of pieces and parts: hardware, software, transmission networks, protocols, languages, very large scale integrated circuits, algorithms, and all components and practices that belong to these. So we could look at computation—or any domain for that matter—as a repository of elements standing ready for particular uses.

We could think of this repository as a toolbox of elements, or functionalities, available for use. But I prefer to think of it as a realm, a world within which certain things can be done. A world entered for what can be accomplished there.

A domain is a realm in the imagination where designers can mentally envisage what can be done—a realm or world of possibilities. Electronics designers know they can amplify signals, shift their frequency, subtract noise from them, modulate carrier waves, set up timing circuits, and make use of a hundred other reliable operations. They think in terms of what can be carried out in electronics world. More than this, if they are expert they are so familiar with their world that they can combine operations and envisage the outcome almost automatically. The physicist Charles Townes, who invented the maser (a microwave forerunner of the laser), had spent years working with operations and devices to do with waves and atomic resonances: the separation of ions via fields, resonating chambers, sensitive high-frequency receivers and detectors, microwave spectroscopy. And he would later use the functionalities of this world in his invention. Experts "lose" themselves in domain worlds; they disappear mentally into them just as we disappear into the world of English when we write a letter. They think in terms of purposes and work these backwards into individual operations in their mental world, much as a composer works a musical theme back into the instrumental parts that will express it.

A domain is also a world in another sense. It is a realm that its users, not necessarily designers, can reach into to carry out mundane tasks, a world where certain manipulations are possible. The procedure for use is always the same. Some object (or activity or business process) is physically entered into a world. Image processing specialists literally enter an image into "digital world" by scanning it into the computer, or photographing it digitally. Once there the object is passed from one operation to another, worked on, transformed, and sometimes combined with other activities and objects within that world. In digital world an image becomes numerical data, so it can be subjected to mathematical manipulations that correspond to color-correction, sharpening, desaturation, distortion to give a wide-lens effect, the addition of a background. When finished being operated on, the object is brought out again in processed form for use in the physical world. The processed image, as manipulated data, is translated back again into a real-world visual image displayed on a computer screen or stored or printed.

Whatever the domain or its world, this maneuver constitutes the domain's real usefulness. We can think of it as entering something "down" into the particular world, manipulating it there in various ways, and hauling it back "up and out" again for use. A live symphony can be brought "down" into electronics world via microphone equipment, manipulated there—processed electronically and recorded, say—and brought back "up and out" again when it enters the physical world again to be played as sound.

What can be accomplished in each domain is distinctive to that domain's world. Some worlds offer a particularly rich set of possibilities. The digital world can manipulate anything that can be reduced to numerical symbols, whether an architectural design or photographic image or the control settings of an aircraft. It offers a vast set of arithmetic operations or sequences of arithmetic and logical steps. And these operations can be executed swiftly: the manipulations work at the ultrafast switching rate of digital circuits.

Other domain worlds are more restricted, more limited in what they can do. Yet they can be extremely effective. Canals in the late 1700s provided a world in which bulk commodities—coal, grain, timber, limestone, and even livestock—could be transported on laden barges cheaply and easily. The physical shipment of bulk goods could be entered into canal world for what could be accomplished there. This meant that goods literally left the domain of roads and dry land and entered a watery world of barge-horses and boatmen and locks and towpaths, a world where things moved slowly to be sure, but where movement was nonetheless fluid and by previous standards effortless. In this world the shipment could be shunted in different directions—switched onto backwaters and side cuttings that stretched across the land. Some of the shipment could be off-loaded part way to make room for further loading. And when a destination was reached the shipment could again enter the domain of roads and horse drayage—the physical world where it started.

Canal world—now a historical one—lacks versatility; it provides really only one functionality, the conveyance of materials, and even this is available only where canals have been dug. But its narrowness is made up for by its cost-effectiveness. Before canals arrived inland transportation was possible only by clumsy ox-drawn carts that lumbered over unpaved roads. As canals became widespread in England in the early 1800s, the cost of transporting coal fell by 85 percent.

What can be accomplished easily in a domain's world constitutes that domain's power. Computation works well as long as you can reduce something to a numerical description for mathematical manipulation. Canal world works as long as you can reduce an activity to something that can be loaded and conveyed by barge. Electronics (the nondigital kind) works well provided the activity somehow can be represented as the movement of electrons. Each of these is good at certain operations. Of course you could in principle accomplish the same task in many different worlds, but the effectiveness would vary. You can sort a customer list easily in digital world. But you could also enter it into electronics world and sort it there. You would somehow have to arrange that different letters would be represented by different voltages, and provide circuits for sensing these voltages and outputting them in order of amplitude. This is doable but awkward. You could even at a stretch enter a customer list into canal world and sort it there. Each barge could be labeled as a customer and pulled forward as the alphabet was slowly called out. This would work, but it would not be the most effective use of canal world.

Different worlds, as I said, offer different possibilities they excel at. Each provides its own set of operations easily accomplished. So it is quite natural that an object or business activity to be worked on be brought into more than one world to make use of what can be accomplished in each. Optical data transmission provides a world—call it *photonics world*—where messages can be sent on optical fiber networks. The work horses here are packets of photons (or light quanta). These carry messages easily and move extremely fast, at close to the speed of light in a vacuum. But there is a problem: unlike electrons, photons carry no electrical charge, they are neutral beings and are therefore difficult to manipulate. And messages fade every few miles—some light in fiber optic cables is always absorbed—so they must be "repeated" or amplified before further transmission. This means the stream of photons must be properly re-formed and resized.

In its early days, photonics world had no direct means to do this, and messages had to constantly exit photonic world and enter electronics world, which could easily realign, amplify, and switch them. But electronics world is a great deal slower; it must rely on electrons moving in response to electrical and magnetic fields, which is by no means instantaneous. So until proper photon amplifiers arrived in the late 1980s (the erbium-doped fiber amplifier, or EDFA), this was like taking a message off the freeway every few miles, putting it on a frontage road for electronic manipulation, then putting it back on the freeway again. The system worked, but the continual leaving and reentering of photonics world was costly and slowed things down.

There is a general lesson here. Costs cumulate anywhere an activity leaves one world and enters another. The shipping of freight containers by sea is not expensive, but transferring freight from the domain of rail haulage into the shipping "container world" requires the cumbersome and expensive technologies of railheads, docks, container handling cranes, and stevedoring. Such "bridging technologies" are usually the most awkward aspect of a domain. They create delays and bottlenecks, and therefore run up costs. But they are necessary because they make the domain available, and control what can enter and leave its world. We can think of a domain as containing a small number of central operations that are streamlined and cheap—maritime container transportation, say. But surrounding these on the outer edges of the domain are the slower and more awkward technologies that allow activities to enter its world and leave it when finished—the docks and gantry cranes of that world. These in general are costly.

I said earlier that domains reflect the power of the worlds they create. But they also reflect its limitations. Architects who domain their design work digitally can produce variations of their ideas almost instantaneously and compute material costs automatically as the act of designing takes place. But the digital domain imposes its own subtle biases on what can be accomplished. Only the quantifiable aspects of the real world can be mapped into the digital world and worked on successfully there. And so, digital architecture can easily yield geometric surfaces that arch and swoop in nice mathematical curves, but as the architectural critic Paul Goldberger says, it has "little patience for funkiness, for casualness, for incompleteness." Surfaces are quantifiable; funkiness is not. Or, I should say, funkiness is not yet quantifiable. If ever it is—if ever you can move a slider on a screen to yield the degree of funkiness you want—then the domain would extend. But for the moment that is not possible. What cannot be accomplished in a world becomes that world's limitation.

I will come back to domains (or technologies in their plural sense) in Chapter 8, to explore how they come into being and develop over time. For the moment let us simply recognize that when we are theorizing about technology we must recognize that this middle layer of technology—these bodies of technology—operate under different rules than do individual technologies. These bodies of technology, or domains, determine what is possible in a given era; they give rise to the characteristic industries of an era; and they provide worlds entered for what engineers can accomplish there.

There is nothing static about these worlds. What can be accomplished constantly changes as a domain evolves and as it expands its base of phenomena. One implication is that innovation is not so much a parade of inventions with subsequent adoptions: the arrival and adoption of computers, or canals, or DNA microarrays. It is a constant re-expressing or redomaining of old tasks—accounting, or transportation, or medical diagnostics—within new worlds of the possible.

5

ENGINEERING AND ITS SOLUTIONS

Until now we have been exploring the nature of technology: what principles it operates under, and what it is in its deepest sense. And we have used these principles to develop a logic of technology, a framework that tells us how technology is structured and operates in the world. Now, in this second part of the book, we will use this logic to explore how technologies come into being and how technology evolves.

But before we start on this, I want to say a few words about something that lies behind much of the reasoning so far. We have been viewing technologies not as stand-alone objects existing en bloc, but as things with inside anatomies. Indeed, once we accept that technologies are constructions—combinations of components and assemblies—we are forced to see them this way. Does this inside view make a difference to how we see technology?

I argue that it does, in two important ways. The first has to do with how technologies modify themselves over their lifetime. If we see technologies from the outside as stand-alone objects, then individual ones—the computer, gene sequencing, the steam engine—appear to be relatively fixed things. They may differ from one version to the next—the computer goes from the Atanasoff-Berry machine, to Eckert and Mauchly's ENIAC, to the EDVAC—and in this way they may change in fits and starts. But when we look from the inside, we see that a technology's interior components are changing all the time, as better parts are substituted, materials improve, methods for construction change, the phenomena the technology is based on are better understood, and new elements become available as its parent domain develops. So a technology is not a fixed thing that produces a few variations or updates from time to time. It is a fluid thing, dynamic, alive, highly configurable, and highly changeable over time.

The second difference lies in how we see technology's possibilities (I am talking now about technology in its collective sense). Looked at from the outside, each technology in the collective appears to fulfill some set of purposes and not much more. If we want to measure we have surveying methods; if we want to navigate, we have global positioning systems. We can do particular things with surveying, and other particular things with GPS. But this is a limited view of what technology is about. Technology does not just offer a set of limited functions, it provides a vocabulary of elements that can be put together—programmed—in endlessly novel ways for endlessly novel purposes.

There is a great difference in looking at technology this way. Imagine a culture at some time in the future where computers have been lost. Its archeologists unearth a 1980s computer, a battered old Macintosh. They hurry it back to their lab, plug it in, and the venerable box flickers to life. They immediately discover ready-made functions they can execute: a MacWrite word-processing program, MacPaint for image-making, an old spreadsheet. These stand-alone functions are useful, they can execute particular tasks and they do this well, and for a long time the investigators use the machine to carry out these separate tasks.

But the Macintosh offers more than this. Deep within it lies the Macintosh Toolbox, a set of internal commands or functions that can be programmed for general purposes. These commands can be combined in certain prescribed ways to create new unthought-of commands and functions. And these new commands themselves can be named and used as components for yet other combinations. After some time of searching out Macintosh lore in old archives, the researchers learn of this and gain access to the internal commands. They discover how to take these individual commands and put them together to do new things, and to use them as building blocks for new commands. At this point the whole enterprise takes off. The researchers have discovered they can program the Macintosh: they can compose within it using a small collection of basic commands that can be combined in endlessly novel ways and can build up complicated commands from simpler ones.

The machine is no longer something that offers a few stand-alone functions. It now offers a language of expression. A world of possibility has opened up.

These two motifs, of technologies adapting by changing their internal parts and of novel structures coming about through fresh combinations, will repeat constantly in this second part of the book. But our main theme, remember, is that technology evolves by combining existing technologies to yield further technologies and by using existing technologies to harness effects that become technologies. How exactly this happens is what I want to explore, particularly in Chapter 9.

As we proceed we will want to keep an eye on two side issues that recur constantly in writings about technology. One is the degree to which Darwin's mechanism applies to technology: to what extent do new "species" in technology arise somehow from variation of old ones and selection of the fittest? The other is to what degree Thomas Kuhn's ideas apply to technology. Kuhn argued that accepted scientific paradigms elaborate over time, encounter anomalies, and are replaced by new ones. Is this also true for technology?

We will also want to keep an eye on something else: innovation. "Innovation" is yet another of those awkward words in technology. Popularly it is invoked wherever some improvement is put into practice or some new idea is tried, no matter how trivial. And it was used by Schumpeter (confusingly, to my taste) to denote the process by which an invention is co-opted into commercial use. I will use the word in its popular sense of novelty in technology. But that novelty, as we will see, comes in several forms: new solutions within given technologies, novel technologies themselves, new bodies of technology, and new elements added to the collective of technology. Rather than explore "innovation" directly—the idea is too diffuse, too nebulous, for that to be useful—I will explore each of these types of novelty or innovation in the chapters to come.

A good place to start this second part of our exploration is with a key question, one important enough that it will take us this chapter and the next to deal with it. Evolution works by new technologies forming from existing ones which act as building blocks. How exactly does this "forming" happen? By what mechanisms do new technologies arise; or more to the point, by what mechanism do building blocks for the creation of new technologies arise? The obvious answer is by some process of radical innovation—invention, if you want to call it that—and we will look at this shortly. But novel building-block elements also arise from standard day-to-day engineering. This may seem surprising at first, so I want to explore in this chapter why that should be so.

To do this, first let me get clear on what I am calling standard day-to-day engineering.

Standard Engineering

What exactly do engineers do in their day-to-day activities? In general, they design and construct artifacts. They also develop methods, build test facilities, and conduct studies to find out how materials will perform or solutions will work in practice. They push forward the understanding of the subjects they deal with, often within specially dedicated institutes and labs. They investigate failures and look for fixes to these. They manage, advise on legal matters, consult, and serve on advisory committees. And they mull over problems—seemingly endless problems—that they talk about, think about, and worry about.

All these activities occupy engineers. But the central one I want to focus on and call *standard engineering* is the carrying out of a new project, the putting together of methods and devices under principles that are known and accepted. Sometimes this is called design and construction, sometimes design and manufacture. Either way it is the planning, testing, and assembly of a new instance of a known technology; not the "invention" of the cable-stayed bridge or the aircraft, but the design and construction of a new *version* of the cable-stayed bridge, say, the Tatara Bridge in Japan; or a new version of the Airbus. For convenience I will simply call this activity "design."

Nearly all design projects are of this kind—the planning and constructing of a new version of a known technology—just as nearly all science activity is the application of known concepts and methods to given problems. But this does not mean standard engineering is simple. There is a spectrum of difficulty: from conventional projects that use customary practices and standard components, to ones that require practices and parts that are experimental, to projects that have a real edge of difficulty and carry some special challenge. In the examples that follow I will set the dial at the more challenging end of this spectrum.

What exactly is involved in standard engineering—in design projects? The basic task is to find a *form*, a set of architected assemblies, to fulfill a set of purposes. This means matching a purpose with some concept of a structure that will meet it, and putting together a combination of assemblies that will bring this structure to reality. This is a process, often a lengthy one. Textbooks usually point to three stages in it. Design proceeds from an overall concept, to the detailed design of the assemblies and parts that will accomplish this, to their manufacture or construction (along with some necessary feedback among these stages). We can call in recursiveness and say that standard engineering proceeds down the hierarchy from an overall concept to individual assemblies, to their subassemblies, and to *their* individual parts, repeating the design process for each of these in turn.

Roughly speaking, things *do* work this way. But only roughly speaking. The process proceeds as much outward from a set of desirable characteristics (or desiderata) as downward in a hierarchy. The purpose itself determines the requirements for the overall concept. This determines the requirements of the central assembly, which determines the requirements for its supporting assemblies, which determine the requirements for *theirs*. "We wanted to design a 350-passenger airplane," says Joseph Sutter, director of engineering for the Boeing 747 in the late 1960s, "and having conceived the wide single deck, we knew that in the tourist section it could go to nine or ten abreast. That pretty much defined the length of the fuselage. The wing we tried to optimize to give the airplane the lifting capabilities, range, and fuel efficiency that we wanted. The [wing] span was first determined by the aerodynamic requirements of getting the takeoff weight into the air, getting to a good initial cruise altitude and ending up with a reasonable approach speed so pilots would have an easy time landing the aircraft."

Notice the sequence here. The requirements start from the key purpose of the aircraft—to carry 350 passengers—and proceed outward, the needs of one assembly determining those of the next. The assemblies at each level must match and support each other.

We could imagine this process working tidily and marching forward step by step as preplanned. And for some projects it does. But at the challenging end of the spectrum the process is rarely tidy. Usually several concepts—overall design ideas—have been proposed, not just one, and some may get as far as testing or even into detailed design before they are deemed unworkable. Even when a concept is selected it must be translated into assemblies and working components, many of which will have to be specially designed. Designers cannot always predict in advance exactly how these will perform. Early versions may show unexpected glitches: they may not work as expected; they may not work at all; or they may use more weight or energy or cost than visualized. And so, fixes must be sought in the shape of better solutions or materials. And because assemblies must mutually balance so that everything works properly together, unexpected deficiencies in one assembly must be offset by adjustments in others. A design is a set of compromises.

Getting things to work requires a great deal of backing and filling of this kind, as ideas, assemblies, and individual parts are tested and balanced, and difficulties are revealed. If a crucial assembly does not work, it may be necessary to start the project afresh. And if the project is particularly complicated—think of the lunar space program—it may be necessary to break it into steps with different experimental versions of the technology acting as markers, each building on what has been learned from its predecessors.

When a project is exploring into unknown territory, glitches come close to being unavoidable. When the Boeing 747 was conceived in 1965, its larger weight dictated a much more powerful engine than had been used until then. This required not just a scaled-up version of the standard turbofan, but moving to one with a much higher bypass ratio (nearly 6:1 instead of 1:1 for the ratio of air blown by the fan to that from the core turbojet). Pratt & Whitney was redesigning its JT9D engine for the purpose, a large turbofan where as much as 77 percent of the thrust came from an eight-foot diameter fan assembly mounted in front of the compressor. The new powerplant promised a leap forward in performance, but its innovative features caused repeated difficulties. Its variable stators (which help control the airflow through the compressor blades) were controlled by movable linkages, but these stuck periodically (the solution was liberal application of WD-40). Its higher combustion temperatures required a better means of turbine cooling.

One of the worst problems turned out to be the way the engine was mounted on the wing. When the engine was run up to high thrust, it would push forward on its mounting and "bend," causing a slight ovalization in its casing. "The problem was that the engine was so big and heavy it bowed on takeoff," says Robert Rosati, deputy program manager on the JT9D. The deflection was not great, it was about four one-hundredths of an inch, but that was enough to cause the high-pressure-compressor blades to rub at the bottom of the casing. This sort of problem is not dangerous; a certain amount of rubbing can be tolerated without damage. But it caused unacceptable losses in efficiency and reliability. Pratt & Whitney attempted several fixes—stiffening the casing, using pre-ovalized seals that could tolerate abrasion—without success. The eventual solution was an inverted Y-frame mounting—essentially a means to transfer the thrust forces to places where they would cause less bending. Deflection fell by 80 percent, tolerable enough for the purpose, but the setback delayed the whole 747 launch. Such frustrations are not unusual with on-the-edge projects like this.

In the days of the 747, producing a new design required calculations done by hand, thousands of them, along with detailed drawings and mockups. Nowadays computers have taken over this work. Computers can convert design ideas almost instantly into detailed drawings and parts requirements, can create virtual mockups, and sometimes can even direct the manufacture of parts. But even with computer-assisted design and manufacture, human input is still important. Getting things to work requires decisions that cannot be left to a machine. Designers must exercise judgment on concepts, architectures, and materials; and on appropriate strengths, ratings, and capacities. And matching all these at the many levels of a project requires human coordination.

The very scale of a project can make coordination challenging. Different assemblies may be designed by different groups—different companies, even—and these need to be balanced. One team's solution may create obstacles for another's, and harmonizing these may require a great deal of iterative discussion. On this scale standard engineering becomes a form of social organization and there is nothing at all tidy about it. The historian Thomas Hughes emphasizes that whether a new project is successful—meaning that it results in a finished viable design—depends to a high degree on the larger network of interests surrounding it: its engineering champions, funding bureaucracies, sponsors and other participants who stand to gain or lose power, security, or prestige from the finished work. Design and development is a very human process of organization and action.

Engineering as Problem Solving

Earlier I said that engineers spend a great deal of their time—sometimes nearly all of their time—solving problems. Why should this be so? Teachers teach and judges judge, so why should engineers not just engineer? Why should they spend so much of their time solving problems?

From what we have seen, encountering bad luck would seem to be the explanation—running into the sort of unexpected setbacks we saw with the 747, together with the frustrations that come with trying to coordinate many parties with different interests in the project. Setbacks and human difficulties of course are important, but they are not quite the main factors. Engineering and problem solving go hand in hand for a more systemic reason.

Standard engineering, as I have defined it, deals with known technologies. This makes each design a new version—a new instance—of something already known (the JT9D is a new instance of the jet engine). But a new instance, a new design project, is only called for if some aspect of the technology needs to be different (if that were not the case, the known standard design could be taken off the shelf and construction would follow). A new level of performance may be required (as with the JT9D); or a different physical environment may have to be designed for; or better performing parts and materials may have become available; or the market may have changed, calling for a new version of the technology. Whichever the case, a new design project is undertaken only when something different must be designed for.

This means that *a new project always poses a new problem.* And the overall response—the finished project—is always a solution: the idea, a specific one, of an appropriate combination of assemblies that will carry out the given task. We can say that a finished design is a particular solution to a particular engineering problem.

Actually, we can say more than that. Because the overall solution must meet new conditions, its chosen assemblies at each level will need to be rethought—and redesigned to conform. Some assemblies and modules of course can be taken off the shelf and modified, but in general when a new version of an existing technology is constructed, each level, and each module at each level, must be rethought and if it does not fit with the rest or work as hoped, it must be redesigned. Each of these redesigns poses its own problem. So more accurately we can say a finished design is a set of solutions to a set of problems.

Saying this does not mean that every solution is satisfactory. A poor solution can cause persistent problems. The Boeing 737 was plagued by rudder malfunctions (anomalies, in the jargon), that caused at least one crash and took considerable time to understand and fix. Often, indeed, a design problem is not quite resolved in the lifetime of a technology. The problem of providing a proper system of controls for a given aircraft has persisted from the Wright Brothers' Flyer to the modern F-35 jet fighter. Fly-by-wire (control by computer) was brought in as a solution to this, but it is a solution that itself progresses and develops with each new aircraft type that uses it.

Combination and Solutions

All this tells us why engineering at the more challenging end of the spectrum is essentially a form of problem solving, and why engineers constantly think about problems.

But this poses a new question. If engineering is about problems, what exactly constitutes a "solution" or set of solutions to a problem? I said casually that a solution is an appropriate combination that carries out a given task. Any creation in engineering is a construction—a combination of elements—for some purpose. So we can re-ask our question this way: How is a solution in engineering a construction and how exactly does this involve combination?

Actually, this is a central issue for us. All through this book we have been thinking of technologies as combinations—which makes design a process of combination. How exactly does combination work in design?

Certainly engineers select appropriate parts and put them together—they combine them—to work jointly. But this does not mean that they set out deliberately to combine anything or are even conscious they are combining things. Engineers see themselves simply as engaged in fulfilling some purpose or meeting certain specifications, and as solving the problems these bring up. The mental part of this of course calls for choices, and the components chosen together form a combination. But combination is not the goal of the creation process in engineering. It is the *result* of these choices, the result of putting elements together for a new instance of a technology. Combination is a byproduct.

There is an analogy to this in the way you express a thought. Modern psychology and philosophy both tell us that the initial part of thinking does not take place in words. We pull up our ideas—the thought—from some unconscious level, *then* find a combination of words and phrases to express them. The thought exists, and its expression in words follows.

You can see this, or I should say feel this, if you speak more than one language. Suppose your company is doing business in Moscow and some of the people around the table with you speak only Russian, some only English. You have something to say and you express that thought in Russian; a moment later you express the same thought in English. The "thought" exists somehow independent of how you put it in words. You have an intention of saying something, and find words by some subconscious process to express it. The result is utterance. It can be short and spontaneous, as in a conversation; or lengthy, put together piece by piece, as with a speech you are preparing. Either way it is a combination of ideas and concepts linked together for some purpose, expressed in sentences and phrases, and ultimately in words. You are not mindful of creating a combination, but you have done that nonetheless.

It is the same with technology. The designer *intends* something, picks a toolbox or language for expression, envisions the concepts and functionalities needed to carry it out in his or her "mind's eye," then finds a suitable combination of components to achieve it. The envisioning can happen at one time more or less spontaneously. Or it can be drawn out, and put together in parts with much revision. We will look at how such creation works in more detail in the next chapter. But for now, notice that as with language, intention comes first and the means to fulfill it—the appropriate combination of components—fall in behind it. Design is expression.

This carries an implication about engineering and its standing as a creative field. Engineering is often held to be less creative than other fields where design is important—architecture, for example, or music. Engineering of course can be mundane, but the same can be said for architecture. The design process in engineering is not different in principle from that in architecture, or fashion, or music for that matter. It is a form of composition, of expression, and as such it is open to all the creativity we associate with these.

The reason engineering is held in less esteem than other creative fields is that unlike music or architecture, the public has not been trained to appreciate a particularly well-executed piece of technology. The computer scientist C. A. R. (Tony) Hoare created the Quicksort algorithm in 1960, a creation of real beauty, but there is no Carnegie Hall for the performance of algorithms to applaud his composition. There is another reason too. Compositions in technology are largely hidden. They tend to lie on the inside—who can see how a cell phone designer has solved a particular problem—hidden within some casing, or within lines of algorithm code, or within some industrial process, and not at all visible to the uninitiated.

Once in a while the creativity is visible. In the first decades of the twentieth century, the Swiss engineer Robert Maillart created a series of bridges that were not new as generic technologies, yet were as innovative as any creation of Le Corbusier or Mies van der Rohe. Built in an era where bridges were embellished with decorations and constructed from heavy masonry, Maillart's bridges appear elegant. Even today they seem daringly modern. Civil engineer David Billington describes Maillart's 1933 Schwandbach bridge near Berne in Switzerland as "one of the two or three most beautiful concrete bridges ever built." It seems not so much to span a ravine as to float over it, an object almost recklessly slender, and one supremely innovative.

Yet as a construction Schwandbach used no new form: it was still only a deck supported by vertical members on top of an arch, a standard form well accepted in Maillart's time. It used no new material: reinforced concrete had been used since the mid-1890s. And it used no new parts. Maillart had accomplished his elegance of expression by almost mundane means. He had learned after much geometric analysis—he was not gifted as a mathematician—to greatly stiffen the bridge deck. This helped distribute heavy loads (trucks at one end) evenly across the whole bridge. Think of a physical model of the bridge in which the supporting arch is a strip of metal held securely between two end blocks and flexed between these. Now put a deck— a flat strip of surface—on top of this, supported by vertical rods fixed to the arch and anchored to the blocks at each end. Loading the deck at one end will press down on the arch on that side and cause it to push upward on the other. If the deck is soft it will give, and all this downward force will undesirably be taken in one place, on the loaded side of the arch. But if the deck is stiff, its firmness will counter the pushing upward on the arch's unloaded side and press downward in reaction. So the load will be distributed much more evenly across the whole bridge.

It was this "solution" that allowed lightness of the arch and deck without sacrificing strength—and therein lies the elegance. The lightness also permitted construction from minimal scaffolding that could be built from the ground up. Maillart had also learned to work fluently in the new medium of reinforced concrete, and this freed him from heavy stone masonry. He had done away with embellishments, no small part of the reason his structures still look modern today. The form was both effective and economical. And innovative. It succeeds however because it is more than these. The separate parts and materials combine to give a flow, a harmony to the whole. The finished object is a piece of technology, but it is very much a piece of art.

I do not want to romanticize standard engineering or its master practitioners. The type of mastery you find in a Maillart bridge does not come out of "genius." More than anything it comes out of an accumulation of knowledge and expertise slowly gathered over years, which is exactly what Maillart possessed. And not all instances of standard technology are in a class with Maillart's. Most projects consist in applying standard solutions to standard problems. The required dimensions or specifications are changed somewhat, but not much more than recalculation and redesign from some standard template are necessary. Still, even the most mundane of projects is a solution to a problem, or a set of solutions to a set of problems, and as such is open to creativity.

This has a consequence. Design is a matter of choosing solutions. Therefore it is a matter of choice. It would seem that the choices are restricted, given that all parts of a technology are tightly constrained by such things as weight, performance, and cost. But constraints more often act to complicate the problem being solved and therefore to call for a larger number of parts to accomplish the job. In fact, for a project of complication the choices of expression—the number of solutions and solutions-within-solutions (subsolutions)—are huge. Any new version of a technology is potentially the source of a vast number of different configurations.

In practice the number of configurations will be fewer than the number possible because engineers tend to repeat the solutions—the phrases and expressions—they have used earlier, and they tend to use off-the-shelf components where they can. So a single practitioner's new projects typically contain little that is novel. But many different designers acting in parallel produce novel solutions: in the concepts used to achieve particular purposes; in the choice of domains; in component combinations; in materials, architectures, and manufacturing techniques. All these cumulate to push an existing technology and its domain forward. In this way, experience with different solutions and subsolutions steadily cumulates and technologies change and improve over time. The result is innovation.

The economic historian Nathan Rosenberg talks about the cumulative impact of such small improvements. "Such modifications are achieved by unspectacular design and engineering activities, but they constitute the substance of much productivity improvement and increased consumer well-being in industrial economies." Standard engineering contributes heavily to innovation.

Standard engineering *learns*.

Solutions Becoming Building Blocks

Standard engineering does something else. It contributes to the evolution of technology. The reader may already have anticipated this from what I have said. Much of the time the solutions to engineering problems are particular and do not join the repertoire of technology as a whole. But occasionally some are used repeatedly enough to become objects in their own right, and they go on to become new elements used in the construction of further technologies.

If you look at any engineering handbook, you will see scores of solutions to standard problems. One such book I possess, *Mechanisms and Mechanical Devices Sourcebook*, shows "nineteen methods for coupling rotating shafts," and "fifteen different cam mechanisms." Another, in electronics, illustrates five types of oscillating circuits:

the Armstrong oscillator, the Colpitts, the Clapp, the Hartley, the Vackar. Such handbooks offer standard solutions to problems that come up repeatedly, designs that can be modified for particular uses. Sometimes these solutions arise as inventions proper, formally sought-after answers to unsolved problems. But more often they arise through practitioners finding a new way, a new clever combination of existing components and methods that resolves a standard problem. If the resulting design is particularly useful, it gets taken up by others, begins to spread through the community, and finds general use. It becomes a new building block.

The process works much as Richard Dawkins's memes do. Memes, as Dawkins originally conceived of them, are units of cultural expression such as beliefs, or catchphrases, or clothing fashions. They are copied and repeated and propagate across societies. Successful solutions and ideas in engineering behave this way. They too are copied and repeated and propagate among practitioners. They become elements that stand ready for use in combination.

In fact, if used often enough, a solution—a successful combination—becomes a module. It gets its own name and becomes encapsulated in a device or method as a module available for standard use. It becomes a technology in itself. There is a parallel to this in language, when a new term summarizes some complicated group of thoughts and becomes a new part of the vocabulary. The words "Watergate" or "Munich" started as constructions that summarized a complicated set of specific government misdoings or negotiations. Now the suffix "-gate" and the word "Munich" have solidified into pieces of vocabulary, available to denote any government misdeed or political appeasement. They have become building blocks in the language, adding to the repertoire of elements available for construction in English.

Is this mechanism of solutions generating building blocks Darwinian? Well, as I have described it, it *sounds* Darwinian. Solutions to engineering problems vary and the better ones are selected and propagate themselves. But we need to be careful here. New solu-

tions do not come about by incremental step-by-step changes, as biological changes are forced to. They are combinations that can be put
together instantly and they come out of purposeful problem solving.

What we *can* say precisely is this. The process of problem solving
in engineering brings forth novel solutions—novel combinations—
in an abrupt way that does not match Darwin's slow cumulation
of changes. Then from these, the better ones are selected, and then
propagate through engineering practice, à la Darwin. Some of these
become elements for the construction of further technologies. The
primary mechanism that generates building blocks is combination;
Darwinian mechanisms kick in later, in the winnowing process by
which only some of these solutions survive.

This winnowing, by the way, does not mean that in technology
the best—or fittest—solutions always survive. When several ways of
solving a given problem in engineering arise, we can think of these as
competing for use, for adoption by engineering designers. As a solution becomes more prevalent, it becomes more visible, and therefore
more likely to be adopted and improved by other designers. Small
chance events—who talked to whom at what time, whose method
got mentioned in a trade journal, who promoted what—may push
one of these forward early on. It therefore gains further adoption by
other designers and may go on to "lock in" practice in its domain.
The solution that comes to dominate of course has to have merit,
but may not necessarily be the best of those competing. It may have
prevailed largely by chance.

This process of chance events, prevalence building further prevalence, and lock-in, is something I have written about extensively
before, so I will not go into further details here. It is enough to say
that technologies (or solutions) that gain prevalence tend to gain further advantage and to lock in, so there is a positive feedback process
at work in the "selection" of technologies.

You can see this positive feedback in the case of nuclear reactor
power station designs. One of the main problems here is the choice
of cooling material that transfers heat from the reactor core to a tur-

bine; another is the choice of moderator, which controls the neutrons' energy level in the core. In the early days of nuclear power, many solutions had been suggested—beer, one engineer said, was the only moderator that had not been tried. Three solutions had been extensively developed: light water (H_2O) for both coolant and moderator; heavy water (D_2O) for both; and gas (helium or carbon dioxide, usually) for the coolant with graphite for the moderator.

Different countries and different companies were experimenting with these. Canada favored heavy water because it had the hydroelectric wherewithal to make it, and the British were experimenting with gas-graphite. But no single solution dominated. The United States had several experimental systems under way. But in particular the U.S. Navy, under Admiral Hyman Rickover, had been developing a nuclear submarine program. Rickover chose light water cooling for his submarines even though sodium would have been more efficient and less bulky, because the idea of a sodium leak in a submarine worried him. Sodium explodes in water and can burst into flames when exposed to air. Besides, engineers had a long history with pressurized water, whereas liquid-sodium systems were more untried.

Then in 1949 the Soviet Union exploded its first atomic bomb. One reaction in the United States was to assert its nuclear superiority by showing it possessed a working nuclear power reactor, any nuclear reactor. So the U.S. Atomic Energy Commission, at Rickover's suggestion, took a reactor intended for an aircraft carrier and redesigned it for a land-based one at Shippingport, Pennsylvania. The new reactor, like its naval predecessor, used light water. Subsequent reactors by Westinghouse and General Electric drew from this design and other light water experience; and this, says historian Mark Hertsgaard, "gave the light water model a head start and momentum that others were never able to match and led the industry to base its commercial future on a reactor design that some experts have subsequently suggested was economically and technically inferior."

By 1986, 81 of the 101 reactors under construction worldwide (outside the Soviet Union) were light water. The light water solution had come to dominate. It got ahead early on by small, almost "chance," events, and went on to dominate future designs. Prevalence winnowed out one of several possible "solutions." But as later studies argued, it was not necessarily the best.

Before I gave much thought to "standard engineering," I did not expect it would contribute greatly to technology's innovation or evolution. But as you can see from this chapter, I now believe otherwise. Every project in standard engineering poses a set of problems and every finished result is a set of solutions to these. The useful solutions build up and diffuse among practitioners. And some go on to become additions to the vocabulary of technology; they go on to become elements or building blocks that can be used in further technologies. Standard engineering contributes much to both innovation and evolution.

In the next chapter I want to move to a different question: how novel technologies proper come into being (as opposed to novel versions of known technologies). How, in other words, invention works. Before I look at this, I want to clear up a point. Many purposed systems—trading conventions, tort laws, trade unions, monetary systems—are novel: they are not versions of something that went before. But they are not always deliberately invented either, so they sit between standard engineering and invention. We need to decide what to do about this category.

We can notice that such systems simply emerge over time in a process similar to the way solutions emerge in standard engineering. Trade unions were not "invented." They emerged from associations of journeymen that grew up in the middle ages—covins and brotherhoods—to provide mutual help. And from some of the early records you can almost see them forming:

The shearmen (who cut the nap of woolen cloth) were "wont to go to all the vadletts within the City of the same trade, and then, by covin and conspiracy between them made, they would order that no one among them should work, or serve his own master, until the said master and his servant, or vadlett, had come to an agreement."

Here we catch trade unions in nascent crystalline form (the inner quote is from the 1300s). Over centuries what started as a social practice grew, solidified, propagated, and took on different forms as circumstances demanded.

This type of nondeliberate coming-into-being is not unusual. But I will not say more about it here. We just need to notice for our argument that novel purposed systems can arise nondeliberately as practices or conventions, solutions to some problem in the economy or society; and if useful they can go on to become components in wider systems. But our main focus is on novel technologies that are deliberately created—inventions. So let us ask how these come into being.

6

THE ORIGIN OF TECHNOLOGIES

The central question Darwin needed to answer for his theory of biological evolution was how novel species arise. The equivalent question for our theory is how radically novel technologies arise. Darwin's solution, as I have said, does not work for technology. The jet engine does not arise from the cumulation of small changes of previous engines favored by natural selection. Nor does it arise by simple combination, throwing existing pieces together in some jumbled fashion mentally or physically. "Add successively as many mail coaches as you please," said Schumpeter, "you will never get a railway thereby." That does not rule out combination, but it means it must take place in some more ordered way than pure randomness.

How then do novel technologies arise?

We are really asking how invention happens. And strangely, given its importance, there is no satisfying answer to this in modern thinking about technology. The last major attempts to theorize about invention were in the 1930s, but the subject fell from fashion in the decades that followed, in no small part because the "creative act" deemed to be at its center was held to be imponderable. And so, today invention occupies a place in technology like that of "mind" or "consciousness" in psychology; people are willing to talk about it but not really to explain what it is. Textbooks mention it, but hurry past it quickly to avoid explaining how it works.

We do know several things about how novel technologies arise. We know, mostly from research in sociology, that novel technologies are shaped by social needs; they come often from experience gained outside the standard domain they apply to; they originate more often in conditions that support risk; they originate better with the exchange of knowledge; and they are catalyzed often by networks of colleagues. No doubt these findings are true. But they do not explain *how* a new technology comes into being any more than pointing to suitable soil explains how a seed comes to germinate.

And so at the very core of where technology is generated, at the very core of the process that over decades generates the structure of the economy and the basis of our wellbeing, lies a mystery. From what process then—from what source—do the devices and methods and products that form the economy originate? That is our question.

What Qualifies as a Novel Technology?

First let us get clear on what makes a technology an invention. What qualifies as a radically new technology, one that departs in some deep sense from those that went before? I will define a radically new technology as one that uses a principle new or different to the purpose in hand. A principle, remember, is the method of something's operation, the base way in which something works.

Does this ring true for what we usually think of as inventions? Consider: In the 1970s computer printing was carried out by line-printers, essentially an electronic typing machine with a set of fixed characters. With the coming of the laser printer, computers printed by directing a laser to "paint" text on a xerographic drum, a different principle. In the 1920s, aircraft were powered by a piston-and-propeller arrangement. With the coming of the turbojet, they were powered by gas turbine engines using reactive thrust, a different principle. In the 1940s, arithmetic calculation was carried out by electromechanical means. With the coming of the computer, it was accomplished by electronic relay circuits, a different principle. In all these cases a new technology came into being—the laser printer, the turbojet, the computer—from a new or different base principle.

A change in principle then separates out invention—the process by which radically novel technologies arise—from standard engineering. It also allows us to draw crucial distinctions between mere improvement and real origination. We can say that the Boeing 747 is a *development* of the 707 technology, not an invention. It improves an existing technology but uses no overall new principle. And we can say that Watt's steam engine is an improvement of Newcomen's. It provides for a new component—a separate condenser—but uses no new principle. (Sometimes improvements are more important commercially than pure originations.) In each case we need only judge whether a new and different base principle is at work for the purpose in hand. This properly allows for gray areas. Was Maillart's stiffened deck an improved component, or a novel principle? It is simultaneously both. Depending on the degree of novelty of the base principle, a continuum exists between standard engineering and radical novelty.

As yet this gives us no actual theory as to how new technologies come into being, but we now have a working criterion for what qualifies as "novel," pointing us on the way.

How then does a novel technology come into existence?

The idea we want to build on is that novel technologies—inventions—use new principles. Recall that a principle is a concept, the idea of some effect or phenomenon in use. So a technology built upon a new principle is actually one built upon a novel or different use of some effect—or effects. This gives us a strong hint about where novel technologies come from. They arise from linking, conceptually and in physical form, the needs of some purpose with an exploitable effect (or set of effects). Invention, we can say, consists in linking a need with some effect to satisfactorily achieve that need. (Of course to qualify as more than standard engineering, the principle or use of this effect must be new to the purpose.)

I find it useful to picture this linkage as a chain. At one end of the chain is the need or purpose to be fulfilled; at the other is the base effect that will be harnessed to meet it. Linking the two is the overall solution, the new principle, or concept of the effect used to accomplish that purpose. But getting the principle to work properly raises challenges, and these call for their own means of solution, usually in the form of the systems and assemblies that make the solution possible. We can think of these as the links of the chain that becomes the overall solution.

But the metaphor is not quite finished. Each of these links in turn has its own task to perform and may therefore face its own challenges; it may require its own sublinks, or sub-subsolutions. The chain, not surprisingly, is recursive: it consists of links—solutions—that call for sublinks—further solutions. And these may require further solutions or inventions of their own. We can think of invention as the process that puts this chain together. It is a process that continues until each problem and subproblem resolves itself into one that can be physically dealt with—until the chain is fully in place.

In the real world this process of linkage varies greatly. Some inventions proceed from individuals working alone, others from teams working separately. Some derive from huge programmatic investment, others from private shoestring effort. Some emerge from years of trial and are marked by a sequence of intermediate versions that did not quite fulfill the goal, others appear whole cloth as if from nothing.

But whatever its variations, invention falls into two broad patterns. It may start from one end of the chain, from a given purpose or need, and find a principle to accomplish this. Or, it may start from the other end, from a phenomenon or effect, usually a newly discovered one, and envisage in it some principle of use. In either pattern the process is not complete until the principle is translated into working parts.

These two patterns overlap a great deal, so there is no need to describe both in detail. I will explore the process of invention mainly when it starts from a perceived need. Later I will say a brief word about the other pattern, where invention starts from a phenomenon.

Finding a Base Principle

Let us suppose, then, that invention starts from a purpose, to find the solution to some perceived need. The need may arise from an economic opportunity, the recognition of a potentially lucrative market perhaps; or from a change in economic circumstances; or from a social challenge; or from a military one.

Often the need arises not from an outside stimulus, but from within technology itself. In the 1920s, aircraft designers realized they could achieve more speed in the thinner air at high altitudes. But at these altitudes reciprocating engines, even when supercharged with air pumped up in pressure, had trouble drawing sufficient oxygen, and propellers had less "bite." Needed was a different principle from the piston-propeller one.

Typically such a need sits for some time with at least some practitioners aware of it, but with none seeing an evident solution. If there were one, standard technology would suffice. The question is therefore by definition challenging. Those that do take up the challenge may encounter the situation as a need to be fulfilled or a limitation to be overcome, but they quickly reduce it to a set of requirements—a technical problem to be solved. The originators of the jet engine (the word "inventor" has connotations of lone eccentrics at work, so I will avoid it), Frank Whittle and Hans von Ohain, were both aware of the limitations of the old piston-and-propeller principle and of the need for a different one. But they re-expressed these as a technical problem—a set of requirements to be met. Whittle sought a power unit that was light and efficient, could compensate for the thin air at high altitudes, and could if possible dispense with the propeller. And von Ohain sought a "steady aerothermodynamic flow process," noting that "the air ducted into such a system could be decelerated prior to reaching any Mach-number-sensitive engine component." The need becomes a well-specified problem.

The problem now comes forward as it were, looking to meet an appropriate solution. The mind (for the moment I will treat the originator as a singular mind, but more usually several minds are at work) becomes fixed on the problem. It scans possibilities that might with further development satisfy the desiderata. This search is conceptual, wide, and often obsessive. Newton commented famously that he had come upon his theory of gravitational orbits "by thinking on it continuously." This continuous thinking allows the subconscious to work, possibly to recall an effect or concept from past experience, and it provides a subconscious alertness so that when a candidate principle or a different way to define the problem suggests itself the whisper at the door is heard.

What is being sought at this stage is not a full design. What is being sought, as I said earlier, is a base concept: the idea of some effect (or combination of effects) in action that will fulfill the requirements of the problem, along with some conception of the means to achieve this.

Each candidate principle, when considered seriously, brings up its own particular difficulties, which pose new conceptual subproblems. The new obstacles narrow and redefine what needs to be solved, as the mind realizes that if a certain piece can be achieved, a larger part of the solution will follow or at least be easier to put in place. The process goes back and forth between levels, testing the feasibility of principles at one level and attempting to deal with the problems these raise at a lower level.

The process here resembles the way a route up an unscaled mountain might be planned. To reach the summit is to solve the problem. And to envision a base principle is to posit a promising overall route or major parts of a route, with a given starting point. On the mountain are patches of obstacles: icefalls, awkward traverses, headwalls, stretches subject to avalanches and falling rock. Each new principle or overall plan of climb meets its own difficult stretches that must be got past. Here recursiveness comes into play, because each obstacle stretch becomes its own subproblem and requires its own solution (or subprinciple or subtechnology, in our case). An overall solution is not achieved until some starting point at the base is connected in a reachable way with the summit. Of course, certain stretches of the mountain may have been climbed before—in our context certain subtechnologies may be available and the solution will be biased toward using these. So the process may be more like stitching together known parts than pioneering a complete route from scratch. The process is in part recursive and the whole becomes a concatenation of parts, a combination of stretches. It forms a plan of advancement, or in our case the envisioning of a technology.

Where do these candidate routes—these principles—come from?

They arise in several ways. Sometimes a principle is borrowed—appropriated from some other purpose or domain that uses it. Whittle, in 1928, mulled through many possibilities: rocket propulsion; reaction propulsion using a rotating nozzle; turbine propulsion using a propeller (a turboprop); and a ducted fan blower (a reaction jet) powered by a piston engine—all the while pondering the subproblems these would raise. Each of these possibilities was borrowed from technologies used for other purposes.

Sometimes a new overall principle is suggested by combining previous concepts. In 1940 the British war effort had been searching for a powerful way to transmit radar microwaves. The physicists John Randall and Henry Boot hit on the principle of the cavity magnetron—a cylindrical electron tube used to generate microwaves for radar purposes using a magnetic field to control the electron flow. To do so they combined the positive aspects of a magnetron (its high power output) and of a klystron tube (a device that used resonant cavities to amplify microwaves).

Sometimes a principle is recalled from the past, or picked up from the remark of a colleague, or suggested by theory. In fact, Randall had by chance encountered an English translation of Hertz's book *Electric Waves* in a bookstore. This had suggested to him the notion of a cylindrical resonant cavity—basically a three-dimensional version of the wire loop resonator Hertz analyzed in his book.

Sometimes the principle—the conceptual solution—is stitched together from existing functionalities, each one solving a subproblem. In 1929, Ernest Lawrence had been searching for some means to accelerate charged particles for high-energy atomic collisions. Particles can be accelerated by an electric field; the various proposals in currency at the time, however, were obstructed by the problem of achieving the extremely high voltages necessary to provide a powerful enough field. Glancing over current periodicals in the university library one night, Lawrence came across an article by the Norwegian engineer Rolf Wideröe. Wideröe's idea was to avoid the high voltage problem by using relatively low alternating current voltage to accelerate particles in repeated jolts. He proposed sending them through a series of tubes laid end to end with small gaps between them. The tubes would be arranged so that the particles would arrive at the gaps just as the AC voltage across these gaps peaked. But this meant that as the particles traveled faster, the tubes would need to get longer. Lawrence saw the beauty of the scheme, but calculated that to achieve the energies he wanted, the line of tubes would stretch well outside his laboratory window. (Modern versions can be up to two miles long.) For Lawrence, Wideröe's idea was not practical.

But Lawrence knew, as did any physicist at the time, that a magnetic field caused charged particles to travel in circular paths. "I asked myself the question . . . might it not be possible to use two electrodes [tubes] over and over again by sending the positive ions [particles] back and forth through the electrodes by some sort of appropriate magnetic field arrangement." In other words he could save space by taking two tubes, bending them to form two halves of a circle separated by gaps, and use a magnetic field to force the particles to circle repeatedly within them. He could then apply Wideröe's well-timed voltages across the gaps and the particles would accelerate each time they crossed from one tube to the other. As they circled they would pick up velocity, spiral out, and eventually be led off for high-energy use.

This principle would become the cyclotron. Notice it was arrived at by transforming the problem from that of achieving extremely high voltages, by Widerøe's subprinciple of using low alternating voltage across a series of tubes, then by Lawrence's subprinciple of using a magnetic field to greatly reduce the space needed. The principle here was constructed from existing pieces—existing functionalities.

In all these cases principles are appropriated from or suggested by that which already exists, be it other devices or methods or theory or functionalities. They are never invented from nothing. At the creative heart of invention lies appropriation, some sort of mental borrowing that comes in the form of a half-conscious suggestion.

Sometimes the principle arrives quickly, with little mental effort. But more usually the overall problem sits at the back of the mind, stymied by some difficulty, with no principle in sight. This situation can continue for months, or even years.

The solution, when it comes, may come abruptly. "The key revelation came in a rush," says Charles Townes, of his insight into what would become the maser. And Whittle says:

> While I was at Whittering, it suddenly occurred to me to substitute a turbine for the piston engine [to drive the compressor]. This change meant that the compressor would have to have a much higher pressure ratio than the one I had visualized for the piston-engined scheme. In short, I was back to the gas turbine, but this time of a type that produced a propelling jet instead of driving a propeller. Once the idea had taken shape, it seemed rather odd that I had taken so long to arrive at

a concept which had become very obvious and of extraordinary simplicity. My calculations satisfied me that it was far superior to my earlier proposals.

The insight comes as a removal of blockage, often stumbled upon, either as an overall principle with a workable combination of subprinciples, or as a subprinciple that clears the way for the main principle to be used. It comes as a moment of connection, always a connection, because it connects a problem with a principle that can handle it. Strangely, for people who report such breakthroughs, the insight arrives whole, as if the subconscious had already put the parts together. And it arrives with a "knowing" that the solution is right—a feeling of its appropriateness, its elegance, its extraordinary simplicity. The insight comes to an individual person, not to a team, for it wells always from an individual subconscious. And it arrives not in the midst of activities or in frenzied thought, but in moments of stillness.

This arrival is not the end of the process, it is merely a marker along the way. The concept must still be translated into a working prototype of a technology before the process is finished. Just as a composer has in mind a main theme but must orchestrate the parts that will express it, so must the originator orchestrate the working parts that will express the main concept.

Embodying the Concept in Physical Form

This new phase of embodying the concept in physical form normally will have been already partially under way. Some components of the device or method usually will have been constructed in experiments, and physical trials of the base concept in action may well have been attempted. So this second phase of the invention process overlaps the first. Bringing the concept to full reality means that a detailed architecture must be worked out; key assemblies must be built, balanced, and constructed; measuring instruments must be appropriated; theoretical calculations must be made. All this needs to be backed by encouragement and finance. Competition helps at this stage. In fact it may be intense if rival groups have glimpsed the principle and are developing it.

Bringing what has largely been an idea to reality raises a number of challenges that may have been mentally foreseen but now must be physically dealt with. Solutions are proposed—and fail; parts do not work; redesigns are necessary; tests must be made. This second phase of invention consists mostly in finding working solutions to challenging subproblems, and it has a great deal of the character of standard engineering.

The challenges can be considerable. Gary Starkweather at Xerox in the late 1960s had been seeking a way to print digital bits directly— any image or text a computer could create—rather than be restricted to the slow awkward line-printers (essentially, large typewriters) that then did the job. He had arrived at the central concept, the idea of using a laser to "paint" an image on a xerographic drum, early on. But to make the concept a working reality, he faced several difficulties. Two in particular called for radical solutions. To make the process commercial he would have to be able to scan a page of written text onto the copier drum in at most a few seconds. And if this was to be achieved with high resolution, he figured that the laser beam would need to be capable of being modulated (switched on and off) to mark black or white dots on the drum at the rate of 50 million times per second. Modulating a laser at this rate was not a solved problem at the time. Further, any laser and lens module would be too heavy—would have too much inertia—to be mechanically swung back and forth thousands of times per second as required to scan lines onto the drum. Both problems needed to be resolved before a working technology could be accomplished.

Starkweather solved the modulation problem by developing a very fast shuttering device using a polarizing filter driven by a piezoelectric cell. He solved the inertia problem ingeniously by keeping the laser module stationary and moving only the laser beam itself, using a rotating multifaceted mirror. Each mirror facet could scan a thin line across the drum as the mirror revolved, much as a lighthouse beam scans the horizon. But this solution brought its own sub-subproblem. Adjacent facets of the mirror, Starkweather calculated, would need to be vertically aligned to an extremely tight tolerance of six arc-seconds; otherwise adjacent scan lines would not be properly offset and the image would be distorted. But the costs of machining to such precise tolerances would be prohibitive. A carefully designed cylindrical lens (Starkweather's main expertise was optics) solved the problem by ensuring that adjacent lines fell close even if the mirror facets were slightly misaligned.

What is striking, when you read Starkweather's accounts of this, are the choices that confronted him. Each subproblem could potentially be solved in several ways. Starkweather was choosing solutions, testing them for workability, and trying to put together a coherent whole from them. He was shifting down the recursive ladder as subproblems and sub-subproblems required originations of their own, and shifting back up as these were resolved or abandoned. Progress here nearly always is slow. It becomes an advance across a broad front as knowledge is gained and subtechnology challenges are successively resolved, pressing always toward a version that works properly.

The first pilot device to do this is an achievement. Even if its initial showings are feeble, the moment nonetheless is precious. In all accounts of origination the moment is remembered where the first crude assembly flickered into life and the principle proved itself. The thing works and a milestone has been passed, to the jubilation of those present. "During a seminar with most of the rest of my students in early April of 1954, Jim Gordon burst in," says Townes (speaking of the maser). "He had skipped the seminar in order to complete a test with open ends. It was working! We stopped the seminar and went to the lab to see the evidence for oscillation and to celebrate."

The initial demonstration may indeed be weak, but with further efforts and ad hoc fixes—and subsequent versions with better components—a robust working version emerges, and the new base principle comes into a semireliable state of being. It has taken physical form. All this takes time—time that tries the patience of backers and supervisors. And time in which the most necessary human ingredient is will, the will to bring the principle to life as a working entity. Now the new device or method becomes a candidate for development, and commercial use. It may, if it is fortunate, enter the economy as an innovation.

Invention, as a process, is now complete.

Invention Proceeding from a Phenomenon

I promised to say a brief word about the second pattern, where invention proceeds from a phenomenon. So let me do this. This pattern is again a linking of an effect in use with a purpose, but now the process starts from the other end—the effect one. Typically someone notices an effect, or posits it theoretically, and this suggests an idea of use—a principle. As with the starting-from-a-need motif, supporting pieces and parts must then be worked out to translate the principle into a working technology.

It would seem that things should be simpler here; a phenomenon may directly suggest a principle of use. And often indeed it does. But just as often the principle is not obvious. In 1928, Alexander Fleming famously noticed the phenomenon that a substance within a mold (spores of *Penicillium notatum*, it turned out) inhibited the growth of a culture of staphylococci bacteria and realized this might have use in treating infection. In retrospect this linkage between effect and use seems obvious. But other people, John Tyndall in 1876 and André Gratia in the 1920s, had noticed the effect before him and had not foreseen a medical use for it. Fleming saw the principle because he had been a doctor in the Great War and was appalled by losses due to battlefield infections, so he was open to see a purpose for what appeared to be a spurious effect.

Even when a principle of use is clearly seen, translating it into a working technology is not necessarily easy. If the effect is novel, which is usually the case, it may be poorly understood, and the tools to work with it may be undeveloped. Transforming the *Penicillium* effect into a working therapy first meant isolating and purifying the active substance in the *Penicillium* mold; then its chemical structure had to be worked out; then its curative powers had to be demonstrated in clinical trials; and finally production methods had to be developed. These harnessing steps called for highly specialized biochemical expertise beyond Fleming's reach, and in the end they were carried out by a team of biochemists led by Howard Florey and Ernst Chain at Oxford's Dunn School of Pathology. Thirteen years passed between Fleming's discovery and penicillin's arrival as a working technology.

I have said that invention starts either from a need or a phenomenon. But the reader might object that many cases start from neither. The Wright brothers, surely, started from neither a need nor a phenomenon. The purpose of self-propelled flight and its two base principles (propulsion via light-weight internal combustion and lift from fixed-wing airfoils) had been accepted years before the Wrights. Actually, this type of case is not at all rare. Very often the base principle of a technology has been seen along with a need for it decades before someone finds a way to translate these into subprinciples and parts that accomplish the job. What the Wrights did was to solve the four key subproblems that hindered these principles from becoming a working technology. By careful experiments and multiple attempts they provided means of control and stability of flight; found wing sections with good lift; constructed a lightweight propulsion system; and developed a high-efficiency propeller. Their 1903 powered flight was not so much a demonstration of "an invention"; it was a marker along a lengthy path trodden by others before them.

Such cases do not constitute a different pattern. They are variations of the two patterns I have described. But here the base principle has either emerged naturally or is somehow obvious. The difficulty is to get the principle to work properly, and sometimes this can require years of effort.

What Lies at the Heart of Invention?

I have given a lengthy account of invention. At its heart lies the act of seeing a suitable solution in action—seeing a suitable principle that will do the job. The rest, allowing some high degree of exaggeration, is standard engineering. Sometimes this principle emerges naturally, as something obvious or easily borrowed. But in most cases it arrives by conscious deliberation; it arises through a process of association—mental association.

How exactly does this mental association take place?

To go back to Lawrence, notice that he does not think of the solution he ended up with: combining an electromagnet with an oscillating radio-frequency electric field between two D-shaped containers. He thinks in terms of how he can use achievable actions and deliverable effects—functionalities—in combination to arrive at a solution. He sees the promise of Wideröe's idea of accelerating particles in small jolts, and solves the resulting space problem by using a magnetic field to cause the particles to circle instead of having to travel hundreds of meters. In other words he associates a problem with a solution by reaching into his store of functionalities and imagining what will happen when certain ones are combined.

In retrospect, Lawrence's insight looks brilliant, but this is largely because the functionalities he uses are not familiar to us. In principle, Lawrence's problem does not differ from the mundane ones we handle in daily life. If I need to get to work when my car is in the repair shop, I might think: I could take the train and from there get a cab; or I could call a friend and get a ride if I were willing to go in early; or I could work from home providing I can clear up some space in my den. I am reaching into my store of everyday functionalities, selecting some to combine, and looking at the subproblems each "solution" brings up. Such reasoning is not a mystery when we see it applied to everyday problems, and the reasoning in invention is not different. It may take place in territory unfamiliar to us, but it takes place in territory perfectly familiar to the inventor. Invention at its core is mental association.

The sort of mental association I have been talking about—Lawrence's sort—uses functionalities. The originator reaches into a store of these and imagines what will happen when certain ones are combined. But sometimes associations form directly from principles themselves. Principles often apply across fields. Or, I should say phenomena echo across fields. Where there are waves—acoustic, oceanic, seismic, radio, light, X-ray, particle—there is interference (two or more waves can produce patterns by superimposing on one another); there is a spectrum of frequencies; there is resonance (where the system oscillates at its natural frequency); there is refraction (where a wave changes direction when it enters a new medium); and there are Doppler effects (perceived changes in frequency if the wave source is moving relative to us). All these provide concepts of use—principles. And these in turn can be borrowed from traditional domains of use and set to work in new ones. So originators thinking of a needed functionality—how can I measure motion? how can I create a steady oscillation at a certain frequency?—associate from some given field and borrow principles from it. Randall borrowed Hertz's wire-loop resonating principle and envisaged it working in three-dimensional form as a cylindrical resonant cavity. When originators need a certain function they can associate back to a principle that produced a corresponding one in a field they know about. At the core of this mechanism—call it *principle transfer*—is seeing an analogy. This is another form of mental association.

By saying that the core of invention is mental association, I am not ruling out imagination and creativity. Far from it. Originators must have the imagination—and creativity—to see the problem as important in the first place, to see it might be solvable, to envisage several solutions, to see the necessary components and architectures for each, and to solve the inevitable sub-problems that come up. But there is nothing unearthly about this type of imagination. What is common to originators is not "genius" or special powers. In fact, I do not believe there is any such thing as genius. Rather it is the possession of a very large quiver of functionalities and principles. Originators are steeped in the practice and theory of the principles or phenomena they will use. Whittle's father was a machinist and inventor, and Whittle was familiar with turbines from an early age.

Originators, however, do not merely master functionalities and use them once and finally in their great creation. What always precedes invention is a lengthy period of accumulating functionalities and of experimenting with them on small problems as five-finger exercises. Often in this period of working with functionalities you can see hints of what originators will use. Five years before his revelation, Charles Townes had argued in a memo that microwave radio "has now been extended to such short wavelengths that it overlaps a region rich in molecular resonances, where quantum mechanical theory and spectroscopic techniques can provide aids to radio engineering." Molecular resonance was exactly what he would use to invent the maser.

You can see this cumulation of functional expertise in what originators take for granted. The biochemist Kary Mullis remarks on the simplicity of his polymerase chain reaction scheme (which makes huge numbers of copies of DNA strands from a particular one in a sample). "It was too easy. . . . Every step involved had been done already." But Mullis's "easy" solution was to "amplify DNA by the repeated reciprocal extension of two primers hybridized to the separate strands of a particular DNA sequence." In lay terms this means finding short stretches of DNA (primers) that flag the beginning and end of the DNA stretch to be copied, and separating the DNA double helix into two separate strands. Once the primers are added, the two strands can build from these (using an enzyme called polymerase) and pick up complementary components to form two new double helixes. Repeating this process again and again multiplies the new double helix copies from 2 to 4 to 8 to 16 . . . indefinitely. This was something easy at that time only to a practitioner with considerable experience of functionalities in working with DNA—something easy to Mullis.

The Pyramid of Causality

I have described invention in this chapter as a micro-process by which an individual (or several) comes up with a novel way of doing things. But it is one that occurs in a context. A novel technology emerges always from a cumulation of previous components and functionalities already in place. We can step back from this observation and view origination with a wider-angle lens by seeing a novel technology as the culmination of a progression of previous devices, inventions, and understandings that led up to the technology in question.

In fact, supporting any novel device or method is a pyramid of causality that leads to it: of other technologies that used the principle in question; of antecedent technologies that contributed to the solution; of supporting principles and components that made the new technology possible; of phenomena once novel that made these in turn possible; of instruments and techniques and manufacturing processes used in the new technology; of previous craft and understanding; of the grammars of the phenomena used and of the principles employed; of the interactions among people at all these levels described.

Particularly important in this pyramid of causality is knowledge—both of the scientific and technical type—that has cumulated over time. This knowledge, as historians Joel Mokyr and Edwin Layton have pointed out, is contained within engineering practice itself, but also within technical universities, learned societies, national academies of science and of engineering, published journals. All these form the all-important substrate from which technologies emerge.

This wider perspective does not negate what I said earlier. The pyramid of causality supports the micro-process of invention much as a logistics system supports an army in battle. In fact, I could have used historical causality to explain inventions in place of the personal approach I took. This would be like explaining the Battle of Waterloo in terms of the histories of the regiments that fought, their military culture, their training and equipment, their previous accomplishments, and their supply lines. These ultimately account for battles won, but normally we focus on the actions and decisions at the sharp end of military engagements where the actual fighting takes place.

Saying that new technologies have a causal history does not imply their appearance is predetermined. Invention is subject to the vagaries and timing of the discovery of new phenomena, of the appearance of new needs, and of the insights arrived at by the individuals who respond to these. Still, the fact that all inventions are supported by a pyramid of causality means that an invention tends to show up when the pieces necessary for it, and the need for it, fall into place.

This rough "readiness" in timing makes it rare for a novel technology to be the work of a single originator. Typically several groups of inventors have envisaged the principle in action at more or less the same time and have made attempts at a working version of it. Such multiple efforts and filling in of key pieces in fact make it difficult to speak of "invention" in the sense of being first. In most cases we can find some vague prior articulation or prior embodiment of the principle, perhaps not well grasped, but prior just the same. And almost as often we find a series of prototypic versions by different workers who borrow from each other, with the device or method improving gradually in effectiveness from crude beginnings as improved subtechnologies are found. The computer is an example. It was not quite "invented." We can say that Claude Shannon saw the base principle of using electronic relay circuits to perform arithmetic operations. Then versions of the principle in action built up, borrowing from each other and adding successive improvements and components. The computer formed in no single step.

Assigning invention in cases like this is difficult, and modern writings on technology recognize this. Says computing pioneer Michael Williams:

> There is no such thing as "first" in any activity associated with human invention. If you add enough adjectives to a description you can always claim your own favorite. For example the ENIAC is often claimed to be the "first electronic, general purpose, large scale, digital computer" and you certainly have to add all those adjectives before you have a correct statement. If you leave any of them off, then machines such as the ABC, the Colossus, Zuse's Z3, and many others (some not even constructed such as Babbage's Analytical Engine) become candidates for being "first."

Williams is correct. The facts that the principle occurs to many, and that there exist different working versions of the principle in action, normally thwart efforts to assign credit for "being first" to a single person or group. If credit for "invention" must be assigned, it should go to the person or team that first had a clear vision of the principle, saw its potential, fought for its acceptance, and brought it fully into satisfactory use. And usually there are several of these.

In fact, even with a single originator, human interaction and informal networks of communication greatly enhance the process I have described. They steep the originator in the lore that has built up around the problem, offer suggestions of principles at work in other domains, and provide equipment and know-how to bring concepts to physical reality. To keep the argument simple, I have pushed these to the background. But we need to note they are often important.

Invention in Science and Mathematics

Does the logical structure I have laid out for invention extend to origination in science and in mathematics? With a few necessary changes, my answer is yes. The reason is that theories whether in science or in mathematics are purposed systems, just as technologies are. They are constructions from component systems that fulfill given purposes, and so the same logic as in technology applies to them.

Let me illustrate this very briefly in science with a case the reader will likely be familiar with. In the year or so after Darwin returned from the voyage of the *Beagle*, he was searching for a theory of speciation—an explanation of how, for example, the different species of finch he had observed on the Galapagos Islands had come into being. From his reading and experience he had put together a cluster of supporting facts and notions, which might help him find supporting principles: That the time scale for evolution corresponded to geological time. That the individual should be the central element in speciation. That variations of traits were in some way heritable. That variations could allow a species to adapt in a slowly changing environment. That habits acquired during the life of the individual might somehow yield inheritable changes. That animal breeders selected for favorable traits they wanted to be inherited. In fact, "I soon perceived that selection was the keystone of man's success in making useful races of animals and plants. But how selection could be applied to organisms living in a state of nature remained for some time a mystery to me." Darwin was struggling to see how these various candidate components might together construct an explanation for speciation.

Then in 1838, "I happened to read for amusement Malthus on Population, and being well prepared to appreciate the struggle for existence which everywhere goes on from long-continued observation of the habits of animals and plants, it at once struck me that under these circumstances favorable variations would tend to be preserved, and unfavorable ones to be destroyed. The result of this would be the formation of new species. Here, then, I had at last got a theory by which to work."

In my language, Darwin did not receive a theory from Malthus. He borrowed a subprinciple: that ongoing competition for scarce resources "selects" the best adapted individuals in a population. And he used this to make workable one of his two main principles, that favorable adaptations would be selected for and could cumulate to produce a new species. The other main principle, that variation generated a range of traits upon which selection could work, he could not reduce to finer explanatory components and had to treat as a postulate. But we can certainly say that in putting the parts together as explanatory functionalities, he had got a theory upon which to work. Arriving at his base principles took him about fifteen months of hard thinking. The rest, the detailed translation of his base principles into a full theory with all the supporting pieces and parts connected to his satisfaction, took him another twenty years.

Origination in scientific theorizing, as in technology, is at bottom a linking—a linking of the observational givens of a problem with a principle (a conceptual insight) that roughly suggests these, and eventually with a complete set of principles that reproduces these.

What about origination in mathematics? This is also a linking, but this time of what needs to be demonstrated—usually a theorem—to certain conceptual forms or principles that will together construct the demonstration. Think of a theorem as a carefully constructed logical argument. It is valid if it can be constructed under accepted logical rules from other valid components of mathematics—other theorems, definitions, and lemmas that form the available parts and assemblies in mathematics.

Typically the mathematician "sees" or struggles to see one or two overarching principles: conceptual ideas that if provable provide the overall route to a solution. To be proved, these must be constructed from other accepted subprinciples or theorems. Each part moves the argument part of the way. Andrew Wiles's proof of Fermat's theorem uses as its base principle a conjecture by the Japanese mathematicians Taniyama and Shimura that connects two main structures he needs, modular forms and elliptic equations.

To prove this conjecture and link the components of the argument, Wiles uses many subprinciples. "You turn to a page and there's a brief appearance of some fundamental theorem by Deligne," says mathematician Kenneth Ribet, "and then you turn to another page and in some incidental way there's a theorem by Hellegouarch—all of these things are just called into play and used for a moment before going on to the next idea." The whole is a concatenation of principles—conceptual ideas—architected together to achieve the purpose. And each component principle, or theorem, derives from some earlier concatenation. Each, as with technology, provides some generic functionality—some key piece of the argument—used in the overall structure.

That origination in science or in mathematics is not fundamentally different from that in technology should not be surprising. The correspondences exist not because science and mathematics are the same as technology. They exist because all three are purposed systems—means to purposes, broadly interpreted—and therefore must follow the same logic. All three are constructed from forms or principles: in the case of technology, conceptual methods; in the case of science, explanatory structures; in the case of mathematics, truth structures consistent with basic axioms. Technology, scientific explanation, and mathematics therefore come into being via similar types of heuristic process—fundamentally a linking between a problem and the forms that will satisfy it.

Invention and Novel Building Blocks

We now have our answer to the key question of how novel technologies arise. The mechanism is certainly not Darwinian; novel species in technology do not arise from the accumulation of small changes. They arise from a process, a human and often lengthy one, of linking a need with a principle (some generic use of an effect) that will satisfy it. This linkage stretches from the need itself to the base phenomenon that will be harnessed to meet it, through supporting solutions and subsolutions. And making it defines a recursive process. The process repeats until each subproblem resolves itself into one that can be physically dealt with. In the end the problem must be solved with pieces—components—that already exist (or pieces that can be created from ones that already exist). To invent something is to find it in what previously exists.

We can now understand why invention varies so much. A particular case can be need-driven or phenomenon-driven; it can have a lone originator or many; its principle may be difficult to conceive of, or may have emerged naturally; translating that principle into physical components may be straightforward or may proceed in steps as crucial subproblems are resolved. But whatever their particular histories, at bottom all inventions share the same mechanism: all link a purpose with a principle that will fulfill it, and all must translate that principle into working parts.

What does this tell us about how novel building blocks form in the collective of technology? Combining this chapter's argument with last chapter's, we can say that novel building blocks arise in three possible ways: as solutions to standard engineering problems (the Armstrong oscillator); as nondeliberate inventions (the monetary system); or as inventions proper, radically novel solutions that use new principles (the jet engine). Whichever the case, all arise from the combination of existing technologies—already existing elements—that provide the necessary functions to make the new element work.

We are not quite finished with individual technologies yet. A new technology is not a fixed thing; it develops. We could say it evolves in the narrow sense of "evolution"; it begins to appear in improved versions and therefore to establish a line of descent. This development, like invention itself, has its own characteristic stages. What are these and what lies behind them?

7

STRUCTURAL DEEPENING

Typically the initial version of a novel technology is crude—in the early days it is sufficient that it work at all. It may have been kludged together from existing assemblies or from parts liberated from other projects; Lawrence's first cyclotron used a kitchen chair, a clothes tree, window glass, sealing wax, and brass fittings. The nascent technology must now be based on proper components, made reliable, improved, scaled up, and applied effectively to different purposes. So its backers construct proper parts and fiddle with the architecture. They test better materials, develop theory, solve problems, and follow dead ends. Slowly and experimentally they eke out improvements and make advances.

The technology begins a journey along a path of development.

This journey in fact will usually have already started. In the process of translating a base concept into physical form, different parts will have been tested, and improvements will have been sought. There is no neat separation between the origination of a technology and its development.

However development has started, once it gets underway, different working versions of the technology emerge. Some of these come from the technology's originators, some from groups of developers attracted into the new field, some possibly from labs or small companies set up to advance the new technology. Each of these will have their own versions of the concept. Still other versions arrive as the new technology also begins to specialize into different forms that suit different purposes and different markets. Radar branched into submarine detection, air and sea navigation, and air traffic control, after it fulfilled its base purpose of detecting aircraft.

All these different versions and branch applications offer different solutions to technical problems. The technology, we can say, shows variation in its solutions. It also shows (or is subject to) selection; over time developers borrow freely from the many available solutions and select some for their designs. This is where Darwinian variation and selection really come in, in technology. The many versions of a technology improve in small steps by the selection of better solutions to their internal design problems.

But designers also improve technologies by deliberate efforts of their own, and invoking Darwin does not tell us how they do this. Nor does it explain a puzzle in the development process. Technologies tend to become more complex—often much more complex—as they mature. The F-35C jet fighter is vastly more complicated than the Wright brothers' Flyer. And to say this has happened by variation and selection from the Wrights' day on challenges the imagination. Something more than pure variation and selection must be going on. Let us look afresh at the development process and see what that is.

Internal Replacement

As a technology becomes a commercial or military proposition, its performance is "pushed": it becomes pressed to deliver more. Its backers seek out better components, improve its architecture, fine-tune and balance its parts in an effort to compete with their rivals. If the competition is severe, even a small edge can pay off handsomely.

But a technology (or more precisely, its base principle) can only be pushed so far before some part of its system runs into a barrier that restrains it. Thus developers press integrated circuits to become denser, therefore accommodating more components; but at some point the process that produces them, photolithography, becomes limited by the wavelength of light. They press radar in its early days to transmit at ever higher frequencies, so as to achieve more accuracy and better discrimination among targets; but at ever higher frequencies, a given transmission source falters in power. A technology can only be pressed so far before it runs into some limitation.

Such obstacles are exasperating to a technology's backers. Each forms a bottleneck that must be taken care of for further progress to happen. But frustrating as this is, reaching such limits is desirable. If a design were not operating close to a limit it would be inefficient, and further performance could be squeezed from it.

Developers can overcome limitations often simply by replacing the impeded component—itself a subtechnology—by one that works better. This might be one that uses a better design, a rethought solution perhaps, or one intellectually appropriated from a rival group. Another way is to use a different material, one that allows more strength per unit of weight, say, or melts at a higher temperature. The jet engine developed over decades by using immensely stronger and more heat resistant alloys for its parts. In fact, often developers are seeking not so much a better part, but a sharper version of the phenomenon that the part provides. And so a great deal of development lies in searching over chemically similar materials for a more effective version of the phenomenon used. Indeed it is safe to say that most of materials science is about seeking improvements in phenomena by understanding the properties of materials that produce these.

The improved component, of course, will require adjustments in other parts to accommodate it; it will require a rebalancing of the technology. It may even require a rethinking of the technology's architecture. When the wooden framing of aircraft was replaced by metal framing in the 1920s and '30s, the whole of aircraft design itself had to be rethought.

What I have said makes it appear that this process of improvement by internal replacement applies to the technology as a whole. But by our recursion principle, it applies to all constituent parts of the technology as well: a technology improves as better subparts and sub-subparts are swapped into its assemblies and subassemblies. This means we need to think of a technology as an object—more an organism, really—that develops through its constituent parts and subparts improving simultaneously at all levels in its hierarchy.

And there is something else. A technology develops not just by the direct efforts applied to it. Many of a technology's parts are shared by other technologies, so a great deal of development happens automatically as components improve in other uses "outside" that technology. For decades, aircraft instruments and control mechanisms benefited from outside progress in electronic components. A technology piggybacks on the external development of its components.

Structural Deepening

Internal replacement gives us part of the explanation for why technologies become more complex as they advance; we could imagine that the parts developers swap in on average are more complicated than the ones they replace. But that is not the main explanation. Another mechanism of development is at work. I will call it *structural deepening*.

Developers can indeed work around an obstacle by finding better components and better materials. But they can also work around it by adding an assembly, or further system of parts, that takes care of it. Here the component presenting an obstacle is not replaced by a different one. It is retained. But additional components and assemblies are added to it to work around its limitation. Thus when jet engines are pushed to perform at higher temperatures and turbine blades begin to soften at these extreme temperatures, developers add a bypass airflow system to cool the blades or a system that circulates a cooling material inside the blades. Or, when the received echoes of early radar systems coincide with the transmitter signal and are drowned out by it, they add a set of parts (together called a duplexer) that turns off the transmitter for a small fraction of a second so that the incoming echo can be received.

By adding subsystems to work around such limitations, technologies elaborate as they evolve. They add "depth" or design sophistication to their structures.

They become more complex.

The forces driving this sort of complication are not just limitations that happen with the technology's overall performance. A technology needs not just to perform well. It needs to be able to operate properly as external conditions change; to be capable of performing a range of tasks; and to be safe and reliable. Limitations can arise with any of these. So we can say that to overcome limits, a technology will add subsystems or assemblies that (a) enhance its basic performance, (b) allow it to monitor and react to changed or exceptional circumstances, (c) adapt it to a wider range of tasks, and (d) enhance its safety and reliability.

This observation applies not just at the overall level of a technology. Its subsystems or assemblies themselves are technologies, and they too develop—are pushed—so as to enhance overall performance. So again, by recursion, the same process applies to them. Designers will be forced to break through *their* limits by adding sub-subsystems according to (a)–(d) above: to enhance performance, react to changed circumstances, adapt to wider tasks, and enhance reliability. The new added assemblies and subsystems in their turn will be pushed, forced to operate at *their* limits of performance. Designers will add further sub-subsystems to break through *these* limits. The process continues with assemblies hung off the main mode to enhance its working, still other subassemblies hung off these to enhance *their* working, and yet other assemblies hung off these. Performance improves at all levels of the system, and at all levels the technology's structure becomes more complicated.

We can see this deepening of structure operating recursively in our example of the gas turbine aircraft engine. Frank Whittle's original prototype had a single compressor that supplied pressurized air for fuel combustion. It was a radial-flow compressor: it took in air and "spun" it fast to compress it. Whittle was familiar with this type of compressor and used it because it was the easiest design to implement. But as better performance was demanded designers began to replace radial-flow compressors with a superior component: the axial flow compressor. Think of this as a giant fan where the airflow is parallel to the driveshaft. A single axial compressor stage however can only supply increased pressure at a ratio of about 1.2:1—a limitation. And so, designers achieved superior air-compression performance by using not one, but several of these. And eventually an assembly of these in sequence. But this compression system needed to operate in both high-altitude thin air and low-altitude dense air and at different airspeeds, and so designers added a guide-vane system to regulate the air admitted. The system was elaborated. The guide-vane system in turn required a control assembly to sense ambient conditions and adjust the vanes accordingly. A further elaboration. But now the outputted high-pressure compressed air could blow backwards through the compressor at times in unexpected pressure surges—a major obstacle. So the compressor system was fitted with an antisurge bleed-valve subsystem to guard against this. Yet more elaboration. The antisurge system in turn required a sensing and control system. Still more elaboration.

In this way a technology (in this case the compressor) improves greatly in its performance and widens the range of environment it can operate in. But at a cost. Over time it becomes encrusted with systems and subassemblies hung onto it to make it work properly, handle exceptions, extend its range of application, and provide redundancy in the event of failure.

In the jet-engine case, other additions improved performance as the technology was pushed. To provide additional thrust in military air-combat conditions, afterburner assemblies were added. To handle the possibilities of engine fires, sophisticated fire-detection systems were added. To prevent the buildup of ice in the intake region, de-icing assemblies were added. Specialized fuel systems, lubrication systems, variable exhaust-nozzle systems, engine-starting systems were added. All these in their turn further required control, sensor, and instrumentation systems and subsystems. Performance was indeed enhanced. Modern aircraft engines are 30 to 50 times more powerful than Whittle's original jet engine, but they are considerably more complicated. Whittle's turbojet prototype of 1936 had one moving compressor-turbine combination and a few hundred parts; its modern equivalent has upwards of 22,000 parts.

This process of improvement by structural deepening is slow. With the aircraft gas-turbine powerplant it took decades. The reason is that not only must the new assemblies and work-arounds be conceived of, they must be tested and proved, and the new system containing them must be rebalanced and optimized. That takes time.

The economics of the process also govern timing. Improvements speed up if competition from other developers becomes heated, and slow down if competition is absent. Developers may be aware of obvious lines of technical advance, but it may not pay to undertake them. Whether competitive pressures are present or not, the set of improvements at any time will still be carefully chosen. Fresh improvements that are technically achievable may be put off until it pays to come out with an overall redesign.

In this way development lurches forward, version by version, as improvements are brought in; and slows as limitations bar the way. The process moves in fits and starts, and it relies heavily on structural deepening.

Structural deepening enables a technology to improve, often considerably. But over time it encrusts the new technology with assemblies and subassemblies needed for superior performance. This may not matter greatly with physical methods and devices. Once development costs have been amortized, the cost may simply be that of materials or space used or weight added. But for other "nontechnological" purposed systems the burden can be considerable. Systems such as military organizations, legal arrangements, university administrations, and word-processing software may also purchase improved performance by adding subsystems and subparts. Think of the steady increase in the complexity of just one legal arrangement, the tax code. But the cost of these "improvements"—in the form of complication and bureaucracy—does not amortize. It is ongoing, and such overhead may be difficult to get rid of when circumstances no longer require it.

Lock-in and Adaptive Stretch

The two mechanisms of development I have described, internal replacement and structural deepening, apply over the life of a technology. Early on, a new technology is developed deliberately and experimentally. Later in its life, it develops more as new instances of it are designed for specific purposes; it becomes part of standard engineering. And, of course, all the while Darwin's mechanism operates. It selects—usually by designers borrowing from their predecessors—the better solutions from the many internal improvements.

But eventually there comes a time when neither component replacement nor structural deepening add much to performance. The technology reaches maturity. If further advancement is sought, a novel principle is needed. But novel principles cannot be counted on to arrive when needed. Even when they do, they may not easily replace the old one. The old design, the old principle, tends to be locked in.

Why? One reason is that elaborations, cumbersome as they are, allow the mature technology to perform better than its nascent rivals. These may offer future potential, but in their infant state they do not perform as well. They cannot directly compete, and so the old technology persists longer than it should.

There is another reason the old principle persists beyond its time, an economic one. Even if a novel principle *is* developed and does perform better than the old, adopting it may mean changing surrounding structures and organizations. This is expensive and for that reason may not happen. In 1955 the economist Marvin Frankel wondered why cotton mills in Lancashire did not adopt the more modern efficient machinery of their American counterparts. He found that the new machinery in the English setting would indeed be more economical. But it was heavy, and to install it the Victorian brick structures that housed the old machinery would have to be torn down. The "outer" assemblies or elaborations thus locked in the inner machinery, and the Lancashire mills did not change.

Still another reason is psychological. The old principle lives on because practitioners are not comfortable with the vision—and promise—of the new. Origination is not just a new way of doing things, but a new way of *seeing* things.

And the new threatens. It threatens to make the old expertise obsolete. Often in fact, some version of the new principle has been already touted or already exists and has been dismissed by standard practitioners, not necessarily because of lack of imagination but because it creates a cognitive dissonance, an emotional mismatch, between the potential of the new and the security of the old. The sociologist Diane Vaughan talks of this psychological dissonance:

> [In the situations we deal with as humans, we use] a frame of reference constructed from integrated sets of assumptions, expectations, and experiences. Everything is perceived on the basis of this framework. The framework becomes self-confirming because, whenever we can, we tend to impose it on experiences and events, creating incidents and relationships that

conform to it. And we tend to ignore, misperceive, or deny events that do not fit it. As a consequence, it generally leads us to what we are looking for. This frame of reference is not easily altered or dismantled, because the way we tend to see the world is intimately linked to how we see and define ourselves in relation to the world. Thus, we have a vested interest in maintaining consistency because our own identity is at risk.

So it is with a novel principle. The greater the distance between a novel solution and the accepted one, the larger is this lock-in to previous tradition. And so a hysteresis—a delayed response to change—exists. The new is delayed by the very success of the old, making changeover in technologies neither easy nor smooth.

This lock-in of an older successful principle causes a phenomenon I will call *adaptive stretch*. When a new circumstance comes along or a demand for a different sphere of application arrives, it is easier to reach for the old technology—the old base principle—and adapt it by "stretching" it to cover the new circumstances.

In practice this means taking the standard components of the old technology and reconfiguring them for the new purpose, or adding further assemblies to them to achieve the new purpose. In the 1930s, high-speed high-altitude flight could have been achieved by the jet engine several years earlier than it was. But designers were not yet familiar with the gas-turbine principle. And so, when pressed to fly military aircraft at speeds that were most achievable in the thinner air at higher altitudes, they adapted and stretched the technology of the day, the aircraft piston engine. This forced piston engines against a limitation. Not only was oxygen scarce at higher altitudes, but the ability to pump oxygen into the cylinder fast enough was itself limited by the rate at which oxygen could be combusted and processed within the four-stroke system. Superchargers and other system deepenings were added to pump in air faster at high pressure. The piston part of the piston-and-propeller principle was elaborated, and with great ingenuity. It was stretched. More difficult to stretch was the propeller. If it worked in the less resistant air of higher altitudes, it would lose bite. If it were pressed to turn at higher revolutions, it would go supersonic. If it were enlarged to have a bigger radius, its tips would travel faster and again go supersonic. A fundamental limitation had been reached.

This is typical. At some point of development, the old principle becomes ever more difficult to stretch. The way is now open for a novel principle to get a footing. The old principle of course lingers, but it becomes specialized for certain purposes. And the new principle begins to elaborate.

The phenomena I am describing in this chapter and the previous one—origination of a new principle, structural deepening, lock-in, and adaptive stretch—therefore have a natural cycle. A new principle arrives, begins development, runs into limitations, and its structure elaborates. The surrounding structure and professional familiarity lock in the principle and its base technology. New purposes and changed circumstances arise and they are accommodated by stretching the locked-in technology. Further elaboration takes place. Eventually the old principle, now highly elaborated, is strained beyond its limits and gives way to a new one. The new base principle is simpler, but in due course it becomes elaborated itself.

Thus the cycle repeats, with bursts of simplicity at times cutting through growing elaboration. Elaboration and simplicity alternate in a slow back and forth dance, with elaboration usually gaining the edge over time.

It will probably have struck the reader that the overall cycle I am describing resembles the cycle Thomas Kuhn proposed for the development of scientific theories. Kuhn's cycle starts when a new theoretical model working on a new principle (he calls this a paradigm) replaces an old one. The new paradigm then is worked out, applied to many exemplars, becomes accepted, and is further elaborated in a process Kuhn calls normal science. Over time, examples that do not fit the base paradigm—anomalies—build up. The paradigm is stretched to accommodate these, but becomes increasingly strained as further anomalies build up. It collapses when and only when a newer, more satisfactory set of explanations—a newer paradigm— arrives.

To show the correspondence I could restate the technology cycle in Kuhn's terms. But it is more interesting to restate Kuhn in the terms I talked of in this chapter. We can say that a theory is "pushed" by being confronted with new facts and by being put to work in new applications. Components of it may need to be replaced—more accurate definitions may be required, and certain constructs may need to be redone. Other aspects when confronted by limitation (anomalies in Kuhn's language) call for special cases— system elaborations that are really workarounds for the perceived limitation. The theory builds out by adding subtheories to handle difficulties and special cases. (Darwinian theory for example must add subarguments to explain why some species show altruistic behavior.)

As the theory develops it elaborates—it adds addenda, further definitions, codicils, and special constructions—all to take into con- sideration different special cases. And if the special cases do not quite fit, the theory becomes stretched; it adds the equivalent of epicycles. Eventually when confronted with sufficient anomalies its "perfor- mance" diminishes and a new principle or paradigm is sought. A novel structure comes into being when the preceding one is stretched and fails. Kuhn's cycle repeats.

This correspondence, let me repeat, does not mean that science is the same as technology. It simply means that because scientific theo- ries are purposed systems, they follow the same logic as technology. They too develop, run into limitations, elaborate, and in due course are replaced. Whether in science or in technology the logic of devel- opment is similar.

We have been exploring the process of development in this chapter. But it is important to remember that it applies to all parts of a technology. A novel technology is "pushed," runs into limitations, and improves through superior parts and structural deepenings. But this same process applies to all of its parts as well. Development is very much an internal process. The whole of a technology and all of its parts develop simultaneously in parallel.

8

REVOLUTIONS AND REDOMAININGS

The last two chapters answered our question of how individual technologies come into being and develop. But what about bodies of technology? How do they come into being and develop? In this chapter I want to explore this parallel question.

It would seem that much of what needs to be said here has already been said. We could think of bodies of technology (or "domains" as I have been calling them) as forming from their constituent technologies and practices, and adding better versions of these as they develop over time. So perhaps we do not need to consider them in a separate chapter here.

But domains are more than the sum of their individual technologies. They are coherent wholes, families of devices, methods, and practices, whose coming into being and development have a character that differs from that of individual technologies. They are not invented; they emerge, crystallizing around a set of phenomena or a novel enabling technology, and building organically from these. They develop not on a time scale measured in years, but on one measured in decades—the digital domain emerged in the 1940s and is still building out. And they are developed not by a single practitioner or a small group of these, but by a wide number of interested parties.

Domains also affect the economy more deeply than do individual technologies. The economy does not react to the coming of the railway locomotive in 1829 or its improvements in the 1830s—not much at any rate. But it does react and change significantly when the domain of technologies that comprise the railways comes along. In fact, one thing I will argue is that the economy does not adopt a new body of technology; it *encounters* it. The economy reacts to the new body's presence, and in doing so changes its activities, its industries, its organizational arrangements—its structures. And if the resulting change in the economy is important enough, we call that a revolution.

How Domains Evolve

Let us dive into the main question of this chapter. How exactly do domains come into being and subsequently develop?

Many, as I said, coalesce around a central technology. As the computer comes into being, supporting technologies—punched card readers, printers, external memory systems, programming languages—begin to gather around it. Others, the grand discipline-based domains in particular, form around families of phenomena and the understandings and practices that go with these. Electronics and radio engineering built from understandings of the workings of electrons and of electromagnetic waves.

Whether domains crystallize around a novel technology or build from a family of phenomena, they are born always from some established field. They must be, because they must construct their original parts and understandings from somewhere—from some parent domain. Computation emerged not from processors or data buses, but out of the components and practices of vacuum-tube electronics in the 1940s.

A new field, at the start of its emergence, is hardly a field at all. It is little more than a cluster of understandings and methods thrown loosely together. It does not offer much, and the existing activities of the economy use it sparingly. They dip into it and select from it as it suits them. Hybrid combinations, drawn from the new domain and its parent one, result. In the early days of railways in Britain, the

Stockton and Darlington Express was a carriage drawn on rails by a horse. Often the new domain's components are enlisted as auxiliaries for an older domain. Early steam engines were used to back up waterwheels in periods of low water, or as pumps to keep the head of water high in a nearby millpond.

At this early stage, the nascent field is still very much part of its parent domain. Genetic engineering began as a minor offshoot of its parent fields of molecular biology and biochemistry. (I am thinking of genetic engineering here in its medical applications: in essence the use of genes to produce or control the proteins necessary to human health.) It grew out of scientific efforts in the middle decades of the twentieth century to understand the mechanisms by which the cell and its DNA manufactured proteins. By the early 1970s, biologists had begun to understand how particular enzymes (restriction enzymes) could cleave or cut DNA at specific places, how others (ligases) could join fragments of DNA together, and how still others could control DNA's replication process. They had begun, that is, to understand the "natural technologies" the cell uses to replicate genes and to direct their production of specific proteins, and slowly they began to capture these and reproduce them artificially in the lab. The biologists had developed techniques for performing tasks that previously only nature could perform.

It is from such component functions that a new domain begins to emerge.

There is an experimental feeling about these early days in any field. But at this stage usually there is little consciousness of a new body of technology emerging. Participants see themselves as solving particular problems in their parent domain. In time though, the new cluster acquires its own vocabulary and its own way of thinking. Its understandings deepen and its practices solidify. Its components improve as each goes through its own version of the development process I talked of in the last chapter. The new domain starts to forget its parentage and strike out on its own, and a consciousness emerges that a new field is starting to exist in its own right.

What has begun to form is a toolbox of technologies and practices. But within this, crucial pieces may function poorly or be missing entirely. These limiting places draw the attention of participants (historian Thomas Hughes calls them "reverse salients," stuck places where the field has been held back despite its advances in neighboring territory), and efforts concentrate there. With sufficient focus, and some measure of good luck, in time the domain breaks through.

If a breakthrough is important it may become an enabling technology: a means to a purpose that promises significant commercial application or offers a key element for the construction of further elements. The technology that brought genetic engineering to life and transformed it from an offshoot of biology to a nascent field of technology was Stanley Cohen and Herbert Boyer's development of recombinant DNA in 1973. Cohen had been researching plasmids, tiny circular units of DNA found in bacteria. And Boyer had been working on restriction enzymes—the "molecular scissors" that can cut DNA into shorter segments. Together the two developed a means to cut a gene from one organism (the frog *Xenopus laevis* in their early experiments) and paste it into a plasmid. They could then insert the plasmid into an *E. coli* bacterium that could reproduce itself rapidly along with its foreign gene. The bacterium would then function as a miniature factory for manufacturing the foreign gene's corresponding protein.

Recombinant DNA was primitive at first, and controversial because of its potential to create nonnatural organisms. But controversial or not, it broke through two limitations that had been holding genetic technology back: the ability to pluck genes from one organism and place them in another within particular organisms, and the ability to manufacture proteins artificially for human purposes.

It is with such enabling technologies that the nascent field reaches adolescence, perhaps even early adulthood. It may now take off. Technical pioneers begin to crowd in and form little companies, a host of improvements follows, and fortunes are founded. An industry starts to grow, the new field becomes exciting, and journalists begin to promote it. Investment may pour in at the prospect of extraordinary profits. Genentech, the first gene-technology company to go public, saw its shares climb from its initial public offering price of $35 to a market price of $89 within twenty minutes of initial trading. Many companies at this stage enter the market with not much more than ideas to offer. Genentech went public in 1980, but offered no real product until it marketed artificial insulin two years later.

The economist Carlota Perez, who has investigated the stages technology revolutions go through, points out that such conditions can bring on an investment mania—and a crash. No crash happened with early gene technology. But crashes have been by no means rare in the past. In Britain in the mid-1840s the wild enthusiasm for railways—"the country is an asylum of railway lunatics," declared Lord Cockburn, a Scottish judge—brought an inevitable collapse. By 1845 a railway mania was in full swing, with scrip (shares that have been sliced into small pieces) sold by alley men, and new schemes for direct lines from little-known towns to other little-known towns launched almost by the day. Then the bubble burst. An economic Week of Terror began on October 16, 1847, in which railway shares lost 85 percent of their peak value, many banks were forced to close, and Britain was brought to the brink of economic collapse. Not all nascent domains bring financial crashes. These happen usually when there is a sense of scramble for space—there are only so many opportunities to build rail lines between Manchester and Liverpool, or between London and Birmingham.

Even if there has been a crash, the new domain survives. And as it further matures it begins to work its way deeply into the economy. Eventually it falls into a stable period of growth. The early competitive frenzy is over and most of the little companies have now vanished; the survivors grow into large corporations. The new period has a different mood. It is one of sobriety and hard work, of confidence and steady growth. The new technology has found its rightful place and has become an underlying part of the economy. This period can last decades and is one of steady buildout of the technology. Railways in Britain expanded from 2,148 miles of track at the peak of the bubble to 21,000 miles sixty-five years later. And during that time they became the engine of growth for the British economy.

As further time passes the domain reaches a comfortable old age. New patents are still granted, but now for less significant ideas, and the once glamorous field no longer inspires. Some domains at this elderly stage are superseded by newer ones, and slowly perish— the canal quietly died as railways came in. But most live on. They become old servants, stalwart retainers, available for duties if needed but largely taken for granted. We still use bridges and roads, sewer systems and electric lighting, and we use them much as we did a hundred years ago. The old dispensations persist and can even muster a tarnished dignity, but we use them without noticing them.

Not all domains go through this cycle of youth, adulthood, and old age quite as neatly as I have described. Some disrupt the cycle by reinventing themselves—changing their character—every few years. They *morph*.

A domain morphs when one of its key technologies undergoes radical change; electronics changed in character when the transistor replaced the vacuum tube. More usually a domain morphs when its main areas of application change. In the 1940s computation was largely an aid to scientific calculation for wartime purposes. By the 1960s, the days of the large mainframes, it was about commerce and accounting. By the 1980s this had given way to office applications on personal computers. And by the 1990s computation was heavily involved in internet services and commerce. Now it is becoming based on what could be called network intelligence. Networks of inexpensive sensors can now see, listen, and pass messages wirelessly to one another. And when attached to trucks, or products on shelves, or machines in factories, these can "converse" and collectively provide intelligent action. Computation, as I write, is morphing again.

Each such morphing grows on top of the previous ones, and each forces the domain's adherents to change their ideas about what the field is all about. But even with such radical changes, the base principles of a field remain the same; computation remains at base the manipulation of objects that can be represented numerically. It is as if the same actor comes onto the stage at intervals in different costumes playing different roles.

Besides morphing, a domain throws off new subdomains. Often these new offspring have multiple parentage. The internet, and the more general domain we call information technology, are both children of computation and telecommunications. They result from the marriage of high-speed data manipulation and high-speed data transmission. The parent fields live on, but they have birthed things that exist on their own, and now a good deal of their energy flows to the new branch. This tendency to morph and to sire new subdomains gives bodies of technology a living quality. They are not fixed collections but miniature ecologies—parts and practices that must fit together at any time, that change constantly as new elements enter, and throw off little subcolonies from time to time that have a different character.

All this means that domains—and by implication any body of technology—are never neatly defined. They add and lose elements; borrow and exchange parts from other domains; and constantly throw off new subdomains. Even when they are relatively well-defined and fixed, their parts and practices vary from one time to another, from one locale to another, and indeed from one country to another. What constitutes computation in Silicon Valley differs from what constitutes computation in Japan.

The Economy Redomaining

What happens to the economy as all this takes place—as bodies of technology emerge and develop?

When it comes to individual technologies, economics is clear. A

new technology—the Bessemer process for producing steel, say—
is adopted, spreads among steel producers, and changes the economy's pattern of goods and services. The Bessemer process made
steel cheaper—much cheaper—than its predecessor, crucible steel,
so steel was used more heavily in the economy. The industries that
used steel—the railroads, construction, heavy machinery—benefited
directly, and this in turn caused *their* costs and what they could offer
their consumers also to change. And the industries that used these
industries in their turn were affected. Such a spreading out of consequences is typical. Just as pulling on one thread of a spider's web
causes the web to stretch and reshape itself in response, so the arrival
of a new technology causes the web of prices and production in the
economy to stretch and reshape itself across all industries.

A similar process applies to bodies of technology. The coming of
the railroad in the 1850s in America not only made transportation
cheaper, it caused readjustments in the parts of the economy that
depended upon transportation. In the Midwest some products—
imported articles and manufactured goods with high transportation
costs that had to be shipped from the East Coast—became cheaper
because of the railroads. Other products such as wheat and hogs
rose in price in the Midwest; their supplies were now drawn off for
shipment to the East. The industries that supplied the railroad were
altered too. Iron production in the United States soared from 38,000
to 180,000 tons in the decade between 1850 and 1860, and the greatly
enlarged output encouraged the adoption of mass-production methods. In turn, changes in the prices and availability of wheat, manufactured goods, and iron affected what was produced and bought in
the myriad of industries that depended upon *these*. So we can say that
bodies of technology cause a pattern of spreading readjustments in
the economy, just as individual technologies do.

This way of looking at things is perfectly valid, but it is not quite
complete. With bodies of technology, something more complicated
than readjustment happens. To see what this is, let me go back to
where the standard argument starts. Economics posits that novel

technologies are "adopted"; they are taken up and used in the economy. This is certainly true for individual technologies. Steelmakers adopt the Bessemer process and their productivity changes accordingly. But this does not describe very well what happens with technologies-plural, such as computation or the railroads. I would prefer to say that the elements of the economy—industries, firms, business practices—do not so much "adopt" a new body of technology; they *encounter* it. And from this encounter, new processes, new technologies, and new industries are born as a result.

How does this happen? Think of the new body of technology as its methods, devices, understandings, and practices. And think of a particular industry as comprised of its organizations and business processes, its production methods and physical equipment. All these are technologies in the wide sense I talked of earlier. These two collections of individual technologies—one from the new domain, and the other from the particular industry—come together and encounter each other. And new combinations form as a result.

Thus, when the banking industry encountered computation in the 1960s, we could loosely say that it "adopted" computation for its bookkeeping and accounting activities. But up close, banking did not merely adopt computation. To be perfectly precise, some activities drawn from bookkeeping (accounting procedures and processes) merged with some activities drawn from computation (particular data-entry procedures, and certain numerical- and text-processing algorithms), and together these formed new functionalities— digitized accounting. The result was a commingling of procedures from banking with procedures from computation that created new processes formed by combinations drawn from both. Such commingling, by the way, is true of all "adoptions." Viewed up close, adoption is always a merging of processes from the adopting field with functionalities from a new domain of possibilities.

In fact, if the new combinations formed this way are powerful enough, they can found a new industry—or at least a subindustry. For decades before banking was computerized, it could design simple options and futures: contracts that clients could purchase allowing them to buy or sell something at a fixed price in the future. Such contracts allowed a farmer planting soybeans in Iowa, say, to sell them in six months' time at the fixed price of $8.40 per bushel, regardless of the market price at that future time. If the price was higher than $8.40 the farmer could sell on the market; if the price was lower he could exercise the option, thus locking in a profit at the cost of purchasing the option contract. The value of the contract "derived" from the actual market value—hence it was called a derivative.

In the 1960s, putting a proper price on derivatives contracts was an unsolved problem. Among brokers it was something of a black art, which meant that neither investors nor banks in practice could use these with confidence. But in 1973 the economists Fischer Black and Myron Scholes solved the mathematical problem of pricing options, and this established a standard the industry could rely on. Shortly after that, the Chicago Board of Trade created an Options Exchange, and the market for derivatives took off.

We cannot quite say here that derivatives trading "adopted" computation. That would trivialize what happened. Serious derivatives trading was *made possible by* computation. At all stages—from the collection and storage of rapidly changing financial data, to the algorithmic calculation of derivatives' prices, to the accounting for trades and contracts—computation was necessary. So more accurately we can say that elements of derivatives trading encountered elements of computation and engendered new procedures, and these comprised digitized derivatives trading. Large-scale trading of derivative products took off as a result.

In fact, even more than this happened. What banking and computation together created in their mutual encounter was more than a new set of activities and products, it was a new set of "programming" possibilities for finance—a new domain of engineering possibilities. In due course financial engineers began to put together combinations of options, swaps (contracts to exchange one cash flow for another), futures, and other basic derivatives for particular purposes such as hedging against future commodities price changes or against foreign exchange fluctuations. A new set of activities had emerged. The encounter between finance and computation had created a new industry, financial risk management.

The process here was one of creative transformation. And it happened over a period of years as problems in the field of financial risk management were solved within the domains of computation and mathematics. The outcome has been an ongoing creative novelty that in finance still has not finished today, and a redomaining of the financial part of banking within the field of computation.

Viewed generally then, the process of redomaining means that industries adapt themselves to a new body of technology, but they do not do this merely by adopting it. They draw from the new body, select what they want, and combine some of their parts with some of the new domain's, sometimes creating subindustries as a result. As this happens the domain of course adapts too. It adds new functionalities that better fit it to the industries that use it.

This overall process in the economy is by no means uniform. It plays out unevenly as different industries, businesses, and organizations encounter the new technology and reconfigure themselves in different ways and at different rates in response. It works outward from changes in the small-scale activities of the economy, to changes in the way business is organized, to changes in institutions, to changes in society itself. A new version of the economy slowly comes into being. The domain and the economy mutually co-adapt and mutually create the new.

It is this process of mutual change and mutual creation that we call a revolution. Each era in the economy is a pattern, a more or less self-consistent set of structures in business and industry and society set in place by the dominant domains of the day. When new bodies of technology—railroads, electrification, mass production, information technology—spread through an economy, old structures fall apart and new ones take their place. Industries that once were taken for granted become obsolete, and new ones come into being.

Old ways of working, old practices, old professions, begin to seem quaint; and the arrangements of work and society become restructured. Many things in the economy remain the same, but many are different forever.

It would be a mistake to say that revolutions happen only when the grand domains change the economy. They also happen as domains of smaller degrees of importance—plastic injection molding, say—come in and cause change at smaller scales. And so, at any time, many revolutions overlap, interact, and simultaneously alter the economy. As these new bodies of technology work into the economy, together they form a mutually consistent set of structures, a roughly consistent pattern in the economy. Each pattern may arise abruptly, lock in for a time, and in time become the infrastructure for the next. Each lays itself down like a geological stratum on top of all that has gone before.

Time in the Economy

All this unfolding of the new technology and readjusting of the economy takes time. A great deal of time. Which explains a puzzle within economics. Usually several decades lie between the coming of enabling technologies that bring a new domain into being and the domain's full impact. The enabling technologies of electrification, the electric motor and generator, arrived in the 1870s, but their full effects on industry were not felt until well into the 1910s and 1920s. James Watt's steam engine was developed in the 1760s, but steam power did not come into prevalent use until the 1820s. In more modern times, the enabling technologies of digitization, the microprocessor and the Arpanet (the forerunner of the Internet), were available by the early 1970s; but again, their impact in digitizing the economy has still not been fully realized. If you accept the adoption story, these delays must be caused by the time people take to find out about the new way of doing things and decide it would improve their circumstances. Such time lags might account for five or ten years. But not three or four decades.

The puzzle dissolves once we admit that what happens is not just a simple process of adoption, but the larger process of mutual adaptation between the domain and the economy I have been talking about. It is not enough that the base technologies of a revolution become available. A revolution does not fully arrive until we organize our activities—our businesses and commercial procedures—around its technologies, and until these technologies adapt themselves to us. For this to happen, the new domain must gather adherents and prestige. It must find purposes and uses. Its central technologies must resolve certain obstacles and fill certain gaps in its set of components. It must develop technologies that support it and bridge it to the technologies that use it. It must understand its base phenomena and develop the theory behind these. Markets must be found, and the existing structures of the economy must be re-architected to make use of the new domain. And the old dispensation must recognize the new domain and become familiar with its inherent practices, which means that practicing engineers who command the grammar of the old need to retool themselves for the new. They do not do this lightly. All this must be mediated by finance, by institutions, by management, by government policies, and by the availability of people skilled in the new domain.

Thus this process is paced not by the time it takes people to notice the different way of doing things and adopt it, but rather by the time it takes existing structures of the economy to re-architect themselves to adapt to the new domain. This time is likely to be decades, not years. And during this time the old technology lives on. It persists despite its demonstrated inferiority.

In 1990 the economic historian Paul David gave a classic example of this process. Before factories were electrified a century or so ago, they were powered by steam engines. Each factory had a single engine, a giant hissing and cranking contraption with pistons and a flywheel and a system of belts and pulleys that turned shafts on several floors of a building. These in turn powered all the factory's machinery. Then electric motors—component technologies of the new electrical domain—became available in the 1880s. They were cheaper in energy use and could be installed as multiple small single units, each next to the machine it powered. And they could be controlled separately, switched on and off as needed. They were a superior technology.

Why then did it take American factories close to 40 years to adopt them? David found that the effective use of the new technology required a different physical construction of the factory than the old steam-engine layout. It required literally that the factory be re-architected. Not only was this expensive, as we saw in the Frankel case, but just how the factory should be constructed was not obvious. The electricians who understood the new domain were not architects, and factory architects were not electricians. So it took considerable time—in this case four decades or more—to accumulate knowledge of how to accommodate factory design to the new technology and for this knowledge to spread. It is not sufficient that businesses and people adapt to a new body of technology. The real gains arrive when the new technology adapts itself to them.

In a sense, these processes of structural change do not just *take* time in the economy, they *define* time in the economy. Let me explain what I mean. Conventionally time is measured as standard clock time. But there is another way to measure time, and this is by the "becoming" of new structures. Philosophers call this relational time. It means that if things stayed always the same, there would be no change by which to mark the passing of things—no change to mark "time." In this sense, time would stand still. By the same principle, if structures changed, if things in the universe were to move and alter themselves, "time" would emerge.

In our case, change—time in the economy—emerges by the underlying structure of the economy altering itself. This happens on two scales, one fast and one slow. The faster scale, call it fast-time, is where the design and testing and absorption into the economy of new individual technologies creates the pace at which things "become," the pace at which new business activities and new ways of doing things form. In conventional time this would be measured in months and years. The slower scale—slow-time—emerges where change is determined by the entry of new bodies of technology. These create eras in the economy and society; they thereby create "time" in the economy. In conventional time this scale would be measured in years and decades.

In both these senses, time does not create the economy. The economy, or changes in its structure, creates time.

Innovation and National Competitiveness

One thing very noticeable about the buildout of new bodies of technology is that their leading edge is highly concentrated in one country or region, or at most a few. Textile technology and steam technology developed largely in Britain in the 1700s; chemical technology developed to a large degree in Germany a century later; and in our own time computation and biotechnology have emerged to a large degree in the United States. Why should this be so? Why should bodies of technology concentrate in particular locations and not spread themselves evenly over many places?

If technology issued forth from knowledge—technical and scientific information—then any country that possessed capable engineers and scientists would in principle be as innovative as any other. Each country, after all, in principle has access to the same science, the same journals—the same knowledge. I do not want to play knowledge down. You cannot advance nanotechnology without detailed knowledge of chemistry and quantum physics. But advanced technology depends on more than knowledge—facts, truths, ideas, and information.

Real advanced technology—on-the-edge sophisticated technology—issues not from knowledge but from something I will call *deep craft*. Deep craft is more than knowledge. It is a set of knowings. Knowing what is likely to work and what not to work. Knowing what methods to use, what principles are likely to succeed, what parameter values to use in a given technique. Knowing whom to talk to down the corridor to get things working, how to fix things that go wrong, what to ignore, what theories to look to. This sort of craft-knowing takes science for granted and mere knowledge for granted. And it derives collectively from a shared culture of beliefs, an unspoken culture of common experience.

It also means knowing how to manipulate newly discovered and poorly understood phenomena, a type of knowing that comes from practical experimentation and research built up in local universities and industrial labs. A knowing that again becomes part of a shared culture. Science too at this level is also craft. In the first three decades of the twentieth century the Cavendish Laboratory at Cambridge was the locus of inventions in atomic physics, and it built these upon a treasury of knowings to do with atomic phenomena. Says the science writer Brian Cathcart, "Whatever was known in this field—techniques, equipment, mathematical tools, even theory—it was known by someone there . . . and more than that it was discussed, challenged and tested at colloquia and other gatherings. To any problem or difficulty in atomic physics there would surely be an answer somewhere in the [Cavendish]."

Such knowings root themselves in local micro-cultures: in particular firms, in particular buildings, along particular corridors. They become highly concentrated in particular localities. There is nothing particularly new in this observation. Here is Alfred Marshall in 1890:

> When an industry has thus chosen a locality for itself, it is likely to stay there long: so great are the advantages which people following the same skilled trade get from near neighborhood to one another. The mysteries of the trade become no mysteries; but are as it were in the air, and children learn many of them unconsciously. Good work is rightly appreciated, inventions and improvements in machinery, in processes and the general organization of the business, have their merits

promptly discussed: if one man starts a new idea, it is taken up by others and combined with suggestions of their own; and thus it becomes the source of further new ideas. And presently subsidiary trades grow up in the neighborhood, supplying it with implements and materials, organizing its traffic, and in many ways conducing to the economy of its material.

Things have not changed since Marshall's day. If anything, the mysteries of the trade are now deeper. This is because they are more likely to be grounded in quantum mechanics, or computation, or molecular biology. These mysteries, or shared knowings, are completely necessary for the processes I have been talking about: invention, development, and the buildout of bodies of technology. They take time to build up, and do not transfer easily to other places. And they cannot be fully written down. Formal versions of the craft do find their way eventually into technical papers and textbooks. But the real expertise resides largely where it was created, taken for granted, shared, and unspoken.

It follows that once a region—or a country for that matter—gets ahead in an advanced body of technology, it tends to get further ahead. Success brings success, so that there are positive feedbacks or increasing returns to regional concentrations of technology. Once a small cluster of firms builds up around a new body of technology it attracts further firms. This is why new bodies of technology cluster in one or two particular regions in a way that becomes difficult to challenge. Other locations can contribute, to be sure. They can manufacture and improve the technology, but they will not be initiating it on a large scale because the detailed knowings needed to push the edge are not resident there.

Of course, regional advantage does not last forever. A region can pioneer some body of technology, but when that ceases to be prominent the region can languish and decline. This can be prevented by parlaying expertise from one body of technology into expertise in another. The region around Stanford University known as Silicon Valley started in wireless telegraphy around 1910, parlayed that into electronics in the 1930s and 40s, seeded a great deal of the computer industry, and has now turned to biotechnology and nanotechnology. As new domains spin off from old ones or sprout from university research, so can a region build upon them.

But countries and regions that are not leaders in technology are not without hope. Well thought out incentives for startup companies and investment in basic undirected science can do much. And technology always spins off from some seed activity, so that if the proper seed is in place, a cluster can build in unexpected places. In the 1980s, the tire industry in Akron, Ohio, was beset by global competition and product recalls, and the big tire companies—B.F. Goodrich, Bridgestone/Firestone, and General Tire—left the region. But Akron possessed from its rubber days a strong expertise in polymer chemistry (the chemistry of chains of molecules), and it managed to lever this into a more general set of knowings, that of high-tech polymer manufacture. Akron is now the center of "Polymer Valley," with more than 400 companies engaged in polymer-related work. Deep knowings in one technology can be levered into deep knowings in another.

All this of course has consequences for national competitiveness. Technology proceeds out of deep understandings of phenomena, and these become embedded as a deep set of shared knowings that resides in people and establishes itself locally—and that grows over time. This is why countries that lead in science lead also in technology. And so if a country wants to lead in advanced technology, it needs to do more than invest in industrial parks or vaguely foster "innovation." It needs to build its basic science without any stated purpose of commercial use. And it needs to culture that science in a stable setting with funding and encouragement, let the science sow itself commercially in small startup companies, allow these nascent ventures to grow and sprout with minimal interference, and allow this science and its commercial applications to seed new revolutions.

The process is not one that can be easily controlled from the top down. There is always a temptation for governments to pursue science with particular commercial aims in view. But this rarely works. Had there been a stated purpose to quantum physics in the 1920s, it would have been deemed a failure. And yet quantum physics has given us the transistor, the laser, the basis of nanotechnology, and much else besides. Building a capacity for advanced technology is not like planning production in a socialist economy, but more like growing a rock garden. Planting, watering, and weeding are more appropriate than five-year plans.

From what I have said in this chapter, it should be clear that domains develop in a different way than individual technologies do. The process is not so much like the development of the jet engine: focused, concentrated, and rational. It is more like the way a system of legal codes forms: slow, organic, and cumulative. With domains, what comes into being is not a new device or method, but a new vocabulary for expression—a new language for "programming" new functionalities in. And this happens by slow emergence. A domain crystallizes around a set of phenomena loosely understood or around a novel enabling technology, and builds organically upon the components, practices, and understandings that support these. And as the new domain arrives, the economy encounters it and alters itself as a result.

All this is another facet of innovation. In fact, we can look on the last four chapters as a detailed explanation of innovation. There is no single mechanism, instead there are four more or less separate ones. Innovation consists in novel solutions being arrived at in standard engineering—the thousands of small advancements and fixes that cumulate to move practice forward. It consists in radically novel technologies being brought into being by the process of invention. It consists in these novel technologies developing by changing their internal parts or adding to them in the process of structural deepening. And it consists in whole bodies of technology emerging, building out over time, and creatively transforming the industries that encounter them.

Each of these types of innovation is important. And each is perfectly tangible. Innovation is not something mysterious. Certainly it is not a matter of vaguely invoking something called "creativity." Innovation is simply the accomplishing of the tasks of the economy by other means.

In the cases I have studied, again and again I am struck that innovation emerges when people are faced by problems—particular, well-specified problems. It arises as solutions to these are conceived of by people steeped in many means—many functionalities—they can combine. It is enhanced by funding that enables this, by training and experience in myriad functionalities, by the existence of special projects and labs devoted to the study of particular problems, and by local cultures that foster deep craft. But it is not the monopoly of a single region, or country, or people. It arises anywhere problems are studied and sufficient background exists in the pieces that will form solutions.

In fact, we can see that innovation has two main themes. One is this constant finding or putting together of new solutions out of existing toolboxes of pieces and practices. The other is industries constantly combining their practices and processes with functionalities drawn from newly arriving toolboxes—new domains. This second theme, like the first, is about the creation of new processes and arrangements, new means to purposes. But it is much more important. This is because a new domain of significance (think of the digital one) is encountered by *all* industries in an economy. As this happens, the domain combines some of its offerings with arrangements native to many industries. The result is new processes and arrangements, new ways of doing things, not just in one area of application but all across the economy.

One last comment. In this chapter and the previous one, I have been talking about technologies "developing": both individual technologies and bodies of technology go through predictable phases as they mature. I could have said instead that each of these "evolves." Certainly they do evolve in the sense that each establishes a line of descent, with all the branching into different "subspecies" or different subdomains that attends this. But I have used "development" because I would rather preserve the word "evolution" for how the whole of technology—the collection of artifacts and methods available to a society—creates new elements from those that already exist and thereby builds out.

This is the central topic of this book. And we have now gathered all the pieces we need to explore it, so let us do that now.

9

———

THE MECHANISMS OF EVOLUTION

Imagine the entire collection of all technologies that have ever existed, past and present. Imagine, that is, all the processes, devices, components, modules, organizational forms, methods, and algorithms in use and ever used. If we were to list these in a catalog their numbers would be vast.

This is the collective of technology, and we want to explore now how it evolves.

I have been claiming that this collective evolves by a process of self-creation: new elements (technologies) are constructed from ones that already exist, and these offer themselves as possible building-block elements for the construction of still further elements. Now I want to make clear the mechanisms by which this happens.

You can see this self-creation of technology in miniature if you look at some small part of this collection building itself. In the early 1900s, Lee de Forest had been experimenting with ways to improve the detection of radio signals, and he had inserted a third electrode in a diode vacuum tube to attempt this. He had been hoping that his triode tube would produce amplification of the signal, something highly desirable given the feebly transmitted radio signals of the day. But it did not. Then, almost simultaneously in 1911 and 1912, several engineers—de Forest among them—did manage to combine the triode with other existing circuit components to produce a workable amplifier. The amplifier circuit together with a slightly different combination of standard components (coils, capacitors, and resistors) yielded an oscillator, a circuit that could generate something highly sought after at the time: pure single-frequency radio waves. This in combination with still other standard components made possible modern radio transmitters and receivers. And these in conjunction with yet other elements made possible radio broadcasting.

And this was not all. In a slightly different circuit combination the triode could be used as a relay: it could act as a switch that could be opened or closed by a small control voltage on the triode's grid. If open, the relay could represent a 0 or the logic value "false;" if closed, a 1 or "true." Relays suitably wired together in combination could yield primitive logic circuits. Logic circuits, again in combination with other logic circuits and electronic elements, made possible early computers. And so, over a period of about four decades the triode vacuum tube became the key building element for a succession of technologies that produced both radio and modern computation.

It is in this way that technology creates itself out of itself. It builds itself piece by piece from the collective of existing technologies. I want to describe the details of how this happens—how technology evolves. How, from so simple a beginning, technology gives us a world of remarkable complexity.

I have been saying casually that technologies are created from existing technologies (or ones that can be created from technologies that already exist). Let me explain why this is true. Any solution to a human need—any novel means to a purpose—can only be made manifest in the physical world using methods and components that already exist in that world. Novel technologies are therefore brought into being—made possible—from some set of existing ones. Always. The jet engine could not have existed without compressors and gas turbines, and without machine tools to manufacture these with the precision required. The polymerase chain reaction was put together from methods to isolate DNA, separate its strands, attach primers, and rebuild double strands from separate ones. It was a combination of things that already existed.

The reader may object that there are exceptions—penicillin seems to be one. It is a therapeutic means and therefore a technology, but it does not seem to be a combination of any previous technologies. But consider: creating a working therapy from Fleming's base effect required a very definite set of existing technologies. It required biochemical processes to isolate the active substance within the mold, other processes to purify it, and still other ones to produce and deliver it. Penicillin had its parentage in these means and methods. It would not have been possible in a society that did not possess such elements. Existing means made penicillin possible. All technologies are birthed from existing technologies in the sense that these in combination directly made them possible.

Of course, the elements that make a technology possible go beyond its mere physical components; they include those necessary in manufacturing or assembling it. And pinning down exact "parentage" may not be simple: the techniques and methods that brought penicillin into existence were many—which should count as parents? The answer of course is the important ones, but which these are is to some degree a matter of taste. Still, this degree of fuzziness does not disturb my central point. All technologies are birthed—made possible—from previous technologies.

Where does this leave us? Strictly speaking, we should say that novel elements are directly *made possible* by existing ones. But more loosely we can say they arise from a set of existing technologies, from a *combination* of existing technologies. It is in this sense that novel elements in the collective of technology are brought into being—made possible—from existing ones, and that technology creates itself out of itself.

Of course, to say that technology creates itself does not imply it has any consciousness, or that it uses humans somehow in some sinister way for its own purposes. The collective of technology builds itself from itself with the agency of human inventors and developers much as a coral reef builds itself from itself from the activities of small organisms. So, providing we bracket human activity and take it as given, we can say that the collective of technology is *self-producing*—that it produces new technology from itself. Or, we can pick up a word coined by Humberto Maturana and Francisco Varela to describe self-producing systems, and say that technology is *autopoietic* ("self-creating," or "self-bringing-forth," in Greek).

Autopoiesis may appear to be an abstract property, the sort of thing that belongs most properly to systems theory or philosophy. But actually, it tells us a lot. It tells us that every novel technology is created from existing ones, and therefore that every technology stands upon a pyramid of others that made it possible in a succession that goes back to the earliest phenomena that humans captured. It tells us that all future technologies will derive from those that now exist (perhaps in no obvious way) because these are the elements that will form further elements that will eventually make these future technologies possible. It tells us that history is important: if technologies had appeared by chance in a different order, the technologies built from them would have been different; technologies are creations of history. And it tells us that the value of a technology lies not merely in what can be done with it but also in what further possibilities it will lead to. The technologist Andy Grove was asked once what the return on investment was for internet commerce. "This is Columbus in the New World," he answered. "What was his return on investment?"

Autopoiesis gives us a sense of technology expanding into the future. It also gives us a way to think of technology in human history. Usually that history is presented as a set of discrete inventions that happened at different times, with some cross influences from one technology to another. What would this history look like if we were to recount it Genesis-style from this self-creating point of view? Here is a thumbnail version.

In the beginning, the first phenomena to be harnessed were available directly in nature. Certain materials flake when chipped: whence bladed tools from flint or obsidian. Heavy objects crush materials when pounded against hard surfaces: whence the grinding of herbs and seeds. Flexible materials when bent store energy: whence bows from deer's antler or saplings. These phenomena, lying on the floor of nature as it were, made possible primitive tools and techniques. These in turn made possible yet others. Fire made possible cooking, the hollowing out of logs for primitive canoes, the firing of pottery. And it opened up other phenomena—that certain ores yield formable metals under high heat: whence weapons, chisels, hoes, and nails. Combinations of elements began to occur: thongs or cords of braided fibers were used to haft metal to wood for axes. Clusters of technology and crafts of practice—dyeing, potting, weaving, mining, metal smithing, boat-building—began to emerge. Wind and water energy were harnessed for power. Combinations of levers, pulleys, cranks, ropes, and toothed gears appeared—early machines—and were used for milling grains, irrigation, construction, and timekeeping. Crafts of practice grew around these technologies; some benefited from experimentation and yielded crude understandings of phenomena and their uses.

In time, these understandings gave way to close observation of phenomena, and the use of these became systematized—here the modern era begins—as the method of science. The chemical, optical, thermodynamic, and electrical phenomena began to be understood and captured using instruments—the thermometer, calorimeter, torsion balance—constructed for precise observation. The large domains of technology came on line: heat engines, industrial chemistry, electricity, electronics. And with these still finer phenomena were captured: X-radiation, radio-wave transmission, coherent light. And with laser optics, radio transmission, and logic circuit elements in a vast array of different combinations, modern telecommunications and computation were born.

In this way, the few became many, and the many became specialized, and the specialized uncovered still further phenomena and made possible the finer and finer use of nature's principles. So that now, with the coming of nanotechnology, captured phenomena can direct captured phenomena to move and place single atoms in materials for further specific uses. All this has issued from the use of natural earthly phenomena. Had we lived in a universe with different phenomena we would have had different technologies. In this way, over a time long-drawn-out by human measures but short by evolutionary ones, the collective that is technology has built out, deepened, specialized, and complicated itself.

What I have said here is brief, and I have not spoken of the mechanisms by which this happens. In the rest of this chapter I want to set out the detailed steps—the actual mechanisms—by which such evolution works. These will comprise the central core of my theory.

Let me do this by talking first about the larger forces that drive technology evolution, then zoom in on its detailed mechanisms. One force certainly is combination, which we can think of as the ability of the existing collective to "supply" new technologies, whether by putting together existing parts and assemblies, or by using them to capture phenomena. The other force is the "demand" for means to fulfill purposes, the need for novel technologies. Together these supply and demand forces bring forth new elements. So let us look at each in turn.

Combination

I have said much about combination already. But just how powerful is it as a potential source of new technologies?

Certainly we can say that as the number of technologies increases, the possibilities for combination also increase. William Ogburn, in fact, had observed this as far back as 1922: "The more there is to invent with, the greater will be the number of inventions." In fact he speculates that the growth of material culture (technologies) shows "resemblance to the compound interest curve." If he were writing today he would say it grows exponentially.

Ogburn gave no theoretical reasoning to back his exponential claim, but we can supply some with a little simple logic. Suppose the collective of technology consists of technologies A, B, C, D, and E only. New workable combinations might include these building blocks in different architectures (think of these as combinations AED and ADE, for example). And they may include them more than once: there might be redundancies (ADDE and ADEEE, for example). But let us be conservative and count possibilities only if they *include* or *do not include* building blocks. No different architectures, that is, and no redundancies. This would give us possibilities of double combinations such as AB, AE, BD; or triple ones like CDE, ABE; or quadruple ones, BCDE, ACDE, and so on.

How many combinations of these types can there be? Well, in a given new combination, each technology A, B, C, D, E may be present or not present. This gives us A or B or C or D or E present or absent—two possibilities for A (present or not), then these times two for B present or not. Counting from A to E this makes two times two times two times two times two, which gives 2^5 or 32 as the number of possibilities. We need to subtract the single technologies where only A, or B, or C is present (these are not combinations), and also the null technology where no building blocks are present. Counting this way, if we have 32 minus the original five, minus the null technology: 26 possibilities. In general, for N possible base elements, we get $2^N - N - 1$ possible technologies. For 10 building-block elements this gives us 1013 combination possibilities, for 20 it gives 1,048,555 possibilities, for 30 it gives 1,073,741,793, and for 40 it gives 1,099,511,627,735. The possible combinations scale exponentially (as 2 to the power of N). For any particular number of building blocks the possible combinations are finite and for small numbers they do not look large. But go beyond these small numbers and they quickly become enormous.

Of course, not all combinations make engineering sense. Many workable possibilities might include processing chips with GPS technology; fewer would include jet engines with hen houses. But even if the chances are 1 in a million that something useful as a building block can be made out of a given set of building blocks, the possibilities still scale as $(2^N - N - 1)/1,000,000$ or approximately 2^{N-20}. They still scale exponentially.

The calculation here is admittedly crude, and we could refine it in several ways. We could allow that many combinations might not make economic sense, which means they might be too expensive for the purposes intended. And some, like the laser or the steam engine, may beget a cascade of further devices and methods issuing from them, while others leave no progeny at all. We could allow that the same components could combine many times over in different architectures. Just one electronic element—the transistor—in combination with copies of itself can create a huge number of logic circuits. The refinements are many. But what I want the reader to notice—and this is my point in doing the exercise—is that even these crude combinatorics demonstrate that if new technologies lead to further new technologies, then once the numbers of elements in the collective pass some rough threshold, the possibilities of combination begin to explode. With relatively few building blocks the possibilities become vast.

Opportunity Niches

Even if new technologies can be potentially "supplied" by the combination of existing ones, they will only come into existence if there exists some need, some "demand" for them. Actually, demand is not a very good word. There was no "demand" in the economy specifically for penicillin or for magnetic resonance imaging before these came into being. So it is better to talk about *opportunities* for technologies—niches they could usefully occupy. The presence of opportunity niches calls novel technologies into existence.

What exactly in human society or in the economy generates opportunity niches?

The obvious answer of course is human needs. We need to be sheltered, fed, transported, kept in good health, clothed, and entertained. There is a temptation to think of these needs as fixed, something like a list of generic want-categories with possible subdivisions of these. But when you delve into any category of need, say the one for shelter, you find it is not fixed; it depends greatly upon the state of society. What we want in shelter—housing—depends greatly on who lives in what, who possesses what, who flaunts what, as a glance at the pages of *Architectural Digest* will confirm. Further, when basic needs are fulfilled and society reaches some degree of prosperity, these "needs" begin to differentiate in an almost vascular expansion. The "need" for entertainment that in early societies was fulfilled by public spectacles and storytelling now requires a panoply of sports, dramas, movies, novels, music. And where the basics in these categories are fulfilled, these needs multiply into subgenres—we have interest in many kinds of music.

There is something else. Human needs are not just created by technology furnishing prosperity, they are created directly by individual technologies themselves. Once we possess the means to diagnose diabetes, we generate a human need—an opportunity niche—for a means to control diabetes. Once we possess rocketry, we experience a need for space exploration.

Like much else in human life our needs are exquisite: they depend delicately and delightfully and intricately upon the state of society, and they elaborate as societies prosper. And because societies prosper as their technologies build out, our needs grow as technology builds out.

But this is still far from the whole picture. The vast majority of niches for technology are created not from human needs, but from the needs of technologies themselves. The reasons are several. For one thing, every technology by its very existence sets up an opportunity for fulfilling its purpose more cheaply or efficiently; and so for every technology there exists always an open opportunity. And for another, every technology requires supporting technologies: to manufacture it, organize for its production and distribution, maintain it, and enhance its performance. And these in turn require their own subsupporting technologies. The automobile in 1900 created a set of ancillary needs—opportunity niches—for assembly-line manufacture, for paved roads and properly refined gasoline, for repair facilities and gas stations. And gasoline in turn set up further needs for refineries, for the importation of crude oil, and for the exploration of oil deposits.

There is a third reason technology generates needs. Technologies often cause problems—indirectly—and this generates needs, or opportunities, for solutions. In the 1600s, mining expanded all across Europe, and as easy-to-access deposits were exhausted, mines ran deeper. Water seepage became a problem, and so a need arose for efficient means of drainage. In 1698 this was fulfilled (not very successfully) by Thomas Savery's "new Invention for Raiseing of Water and occasioning Motion to all Sorts of Mill Work by the Impellent Force of Fire"—a primitive version of the steam engine.

For our argument we do not need a complete theory of how human and technical needs form. But we do need the awareness that the system consists not just of technologies creating technologies, but also of technologies creating opportunity niches that call forth technologies. We also need the awareness that opportunity niches for technologies are not fixed and given, but to a very large degree are generated by technologies themselves. Opportunity niches change as the collective of technology changes; and they elaborate and grow in numbers as the collective grows.

The Core Mechanism

These driving forces give us a broad picture of how technology builds out. Existing technologies used in combination provide the possibilities of novel technologies: the potential supply of them. And human and technical needs create opportunity niches: the demand for them. As new technologies are brought in, new opportunities appear for further harnessings and further combinings. The whole bootstraps its way upward.

By what exact steps—what exact mechanism—does this boot-strapping work?

Think of the collective as a network that builds upon itself and organically grows outward. In this network each technology (I will usually call it an element) is represented by a point or node. Each node has links (directed arrows) pointing to it from its parent nodes, the technologies that made it possible. Of course, not all technologies are actively used in the economy at a given time. We can imagine the elements or nodes that *are* active to be lit up. I will call these the *active collection* of technologies: those elements that are economically viable and are currently used in the economy. The others—the waterwheels and sailing ships of past eras—are essentially dead. They have disappeared from the active collection. They may be revived for use in novel combinations, but this rarely happens.

From time to time new technologies are added to the active set. But not uniformly. At any time the active network grows rapidly in some places and not at all in others. Some elements, usually ones following from recently captured phenomena (such as the laser in 1960), are birthing further elements rapidly. Others, the mature and established ones such as the Solvay process, which produces sodium carbonate, are birthing no offspring. The active network builds out unevenly.

As elements add to or disappear from the active collection of technologies, the collection of opportunity niches changes too. We can imagine these needs as posted in the background on a gigantic bulletin board. (We can think of engineers and entrepreneurs watching this bulletin board and reacting to it.) Each novel element must satisfy at least one need or purpose on the board. As novel elements join the network they may render obsolete elements that previously satisfied their purpose, or elements that cease to be economic. Their opportunity niches also disappear from the bulletin board. Mediating all this is the economy. We can think of it as a system that determines costs and prices and therefore signals opportunities to be fulfilled by novel elements, as well as deciding which candidate technologies will enter the active collection. (For the moment I will take these actions of the economy as given and treat the economy as a black box. I will have more to say about this in Chapter 10.)

Let us ask how the active network of technologies builds out. It evolves over time by a series of encounters between new technological possibilities and current opportunity niches. Each encounter is both an engineering one and an economic one. A candidate solution must technically "work" to be considered for the purpose at hand; and its cost must be in line with what the market is willing to pay for fulfilling the purpose in question. Technologies that fulfill these conditions are potential "solutions" for the purpose at hand. There may be several of them, from which one eventually joins the collective.

We are beginning to get a sense not so much of the steady cumulation of technology, but of a process of the formation of novel elements and opportunity niches and of their replacement and disappearance. This process is algorithmic: it operates by discrete steps.

Let us look at these. We can start by supposing that a candidate novel technology appears. It has been made possible by a combination of previous technologies and has bested its rivals for entry into the economy. Six events or steps then follow. We can think of these as the legitimate moves that can be made in the technology buildout game. I will state them abstractly, but the reader might find it helpful to have an example technology in mind, the transistor say.

1. The novel technology enters the active collection as a novel element. It becomes a new node in the active collection.

2. The novel element becomes available to replace existing technologies and components in existing technologies.

3. The novel element sets up further "needs" or opportunity niches for supporting technologies and organizational arrangements.

4. If old displaced technologies fade from the collective, *their* ancillary needs are dropped. The opportunity niches they provide disappear with them, and the elements that in turn fill these may become inactive.

5. The novel element becomes available as a potential component in further technologies—further elements.

6. The economy—the pattern of goods and services produced and consumed—readjusts to these steps. Costs and prices (and therefore incentives for novel technologies) change accordingly.

Thus the transistor entered the collective around 1950 (step 1); replaced the vacuum tube in most applications (step 2); set up needs for the fabrication of silicon devices (step 3); caused the vacuum-tube industry to wither (step 4); became a key component of many electronic devices (step 5); and caused prices and incentives for electronic equipment to change (step 6).

But listing events this way makes them look too neatly sequential. In practice, they do not follow each other in a tidy way. Often they operate in parallel. A new technology becomes available as a potential building block (step 5)—again think of the transistor—just as soon as it appears (step 1). And new opportunities (step 3) appear almost as soon as a new technology appears (step 1). And of course any of these events takes time to play out. A technology takes time to diffuse through the economy, and the economy in turn may take several years to adjust itself to the novel technology.

If these steps happened one at a time in sequence, the buildout process would be fairly methodical. Each new possibility would add an element and the other five steps would duly follow. But something more interesting happens. Each of these events or steps may trigger a cascade of other events. A novel technology may cause a sequence of further additions to the collective of technology (by steps 3 and 5). Price readjustment (step 6) may cause a sidelined candidate technology suddenly to become viable and enter the active set (step 1). So these steps themselves may trigger a new round of additions to the collective. The appearance of a new technology may set in motion events that never end.

Here is another possibility. A new element may cause not just the collapse of the technology it replaces (step 2), but also the collapse of technologies that depend on the replaced technology's needs (step 4). And as these inferior elements become replaced, *their* dependent opportunity niches also collapse (steps 2 and 4) and with them the technologies that had occupied these. The arrival of the automobile in the early 1900s caused the replacement of horse transportation. The death of horse transportation eliminated the needs for blacksmithing and carriage making. The collapse of blacksmithing in turn eliminated the need for anvil making. Collapses caused further collapses in a backward succession. This is not quite the same as Schumpeter's "gales of creative destruction," where novel technologies wipe out particular businesses and industries broadly across the economy. Rather, it is a chain of domino-like collapses—*avalanches* of destruction, if you prefer to call them that.

The creative side to this is, as Schumpeter pointed out, that new technologies and industries take the place of those that collapse. We can add to this that new technologies can as easily set up new opportunity niches to be occupied by further new technologies, which set up further niches, to be occupied by yet further technologies. There are also avalanches—should we call them *winds*—of opportunity creation.

All this activity is going on at many points in the network at the same time. Like the buildout of species in the biosphere, it is a parallel process, and there is nothing orderly about it.

What I have described is an abstract series of steps, an algorithm, for the evolution of technology. If we start with a few primitive technologies and set this system in motion in our minds, what would we see? Do we see anything like the historical description of technology's evolution I gave earlier?

Well, if we allow the algorithm to play out, at first progress is slow. Not only are technologies few, but because of this, opportunities are also few. Once in a long while a purpose will be satisfied by harnessing some simple phenomenon—in our own historical case the use of fire, or of certain vines for binding. But these primitive technologies offer opportunities, at the very least for better performance of their task. As opportunities are matched, other primitive technologies arrive, perhaps replacing existing technologies. The stock of technologies builds, and with it the stock of available building blocks. And the stock of opportunity niches these bring with them also builds. New combinations or melds of technologies begin to be possible. And as new building blocks form, the possibilities for still further combinations expand. The buildout now becomes busy. Combinations begin to be formed from combinations, so that what was once simple becomes complex. And combinations replace components in other combinations. Opportunity niches begin to multiply; bursts of accretion begin to ripple through the system as new combinations create further new combinations; and avalanches of destruction begin to cascade as replaced combinations take with them their niches for supporting technologies, which causes the further disappearance of the technologies that fulfill these, and their supporting niches. These avalanches vary in size and duration: a few are large, most are small. The overall collective of technology always increases. But the active set varies in size, showing, we would expect, a net increase over time.

There is no reason that such evolution, once in motion, should end.

An Experiment in Evolution

The picture I have just given is imaginary; it is a thought experiment of how the mechanisms of technology's evolution should work in action. It would be better if we could enact these steps somehow in the lab, or on a computer—better to have some model of technology evolve so that we could observe this evolution in reality.

That would be difficult. Technologies differ greatly in type, and it would be hard for a computer to figure out whether some combination of, say, papermaking and the Haber process would make sense and do something useful. But we might be able to confine ourselves to some restricted world of technologies, some world that would evolve on a computer that we could study.

Recently my colleague Wolfgang Polak and I set up an artificial world (one represented within the computer) to do just that. In our model world the technologies were logic circuits. For readers not familiar with these, let me say a word about them.

Think of logic circuits as miniature electronic chips with input and output pins. Inputs to a given circuit might be numbers in the binary form of 1s and 0s. Or they might be some combination of *true* and *false* representing some set of circumstances that are currently fulfilled. Thus the inputs to a logic circuit in an aircraft might check which of engine conditions *A, C, D, H, K,* are *true* or *false,* representing the status say of fuel conditions, or temperatures, or pressures; and the output pins might signal whether certain switches *Z, T, W,* and *R* should be "on" (true) or "off" (false) to control the engine accordingly. Circuits differ in what they do, but for each set of input values a given circuit arranges that a particular set of output values appears on the output pins. The interesting circuits for computation correspond to operations in arithmetic: addition, say, where the output values are the correct summations of the inputted ones. Or they correspond to operations in logic, such as 3-bit AND (if input pins *1, 2,* and *3,* all show *true,* the output pin signals *true;* otherwise it signals *false*).

Working with logic circuits gave Polak and me two advantages. The precise function of a logic circuit is always known; if we know how a logic circuit is wired together, we can figure out (or the computer can) exactly what it does. And if the computer combines two logic circuits—wires them together so that the outputs of one become the inputs of the other—this gives us another logic circuit whose precise function we also know. So we always know how combinations perform, and whether they do something useful.

Polak and I imagined our artificial world within the computer to be peopled by little logicians and accountants, anxious to tally and compare things within this logic-world. At the beginning they have no means to do this, but they have a lengthy wish-list of needs for particular logical functionalities. They would like to have circuits that could perform AND operations, Exclusive-ORs, 3-bit addition, 4-bit EQUALS, and the like. (To keep things simple we imagined this long need list or opportunity-niche list to be unchanging.) The purpose of our computer experiment was to see if the system could evolve technologies—logic circuits—by combination from existing ones to fulfill niches on the list, and to study this evolution as it happens.

At the start of our experiment, as I said, none of these opportunity niches was satisfied. All that was available by the way of technology was a NAND (Not AND) circuit (think of this as a primitive circuit element, a computer chip not much more complicated than a few transistors). And at each step in the experiment, new circuits could be created by combining existing ones—wiring them together randomly in different configurations. (At the start these were simply the NAND ones.) Most new random combinations of course would fail to meet any needs, but once in a long while a combination might result by chance that matched one of the listed needs. The computer was instructed then to encapsulate this as a new technology itself, a new building block element. It then became available as a building-block element for further wiring and combination.

This experiment in technology evolution ran by itself within Polak's computer; there was no human intervention once we pushed the return button to start it. And of course it could be repeated again and again to compare what happened in different runs.

What did we find? In the beginning there was only the NAND technology. But after a few tens and hundreds of combination steps, logic circuits that fulfilled simple needs started to appear. These became building block elements for further combination, and using these, technologies that met more complicated needs began to appear. After about a quarter of a million steps (or 20 hours of machine time) we stopped the evolution and examined the results.

We found that after sufficient time, the system evolved quite complicated circuits: an 8-way-Exclusive OR, 8-way AND, 4-bit EQUALS, among other logic functions. In several of the runs the system evolved an 8-bit adder, the basis of a simple calculator. This may seem not particularly remarkable, but actually it is striking. An 8-bit adder has sixteen input pins (eight each for the two numbers being added) and 9 outputs (eight for the result and one extra for the carry digit). If again you do some simple combinatorics, it turns out there are over $10^{177,554}$ possible circuits that have 16 inputs and 9 outputs, and only one of these adds correctly. $10^{177,554}$ is a very large number. It is far far larger than the number of elementary particles in the universe. In fact, if I were to write it down as a number, it would take up nearly half the pages of this book. So the chances of such a circuit being discovered by random combination in 250,000 steps is negligible. If you did not know the process by which this evolution worked, and opened up the computer at the end of the experiment to find it had evolved a correctly functioning 8-bit adder against such extremely long odds, you would be rightly surprised that anything so complicated had appeared. You might have to assume an intelligent designer within the machine.

The reason our process could arrive at complicated circuits like this is because it created a series of stepping-stone technologies first. It could create circuits to satisfy simpler needs and use them as building blocks to create circuits of intermediate complexity. It could then use these to create more complicated circuits, bootstrapping its way forward toward satisfying complex needs. The more complicated circuits can only be constructed once the simpler ones are in place. We found that when we took away the intermediate needs that called for these stepping-stone technologies, complex needs went unfulfilled.

This suggests that in the real world, radar might not have developed without radio—and without the need for radio communication. There is a parallel observation in biology. Complex organismal features such as the human eye cannot appear without intermediate structures (say, the ability to distinguish light from dark) and the "needs" or uses for these intermediate structures (a usefulness to distinguishing light from dark).

We found other things too. When we examined the detailed history of the evolution, we found large gaps of time in which little happened at all. Then we saw the sudden appearance of a key circuit (an enabling technology) and quick use of this for further technologies. A full adder circuit might appear after say 32,000 steps; and 2-, 3-, and 4-bit adders fairly quickly after that. In other words, we found periods of quiescence, followed by miniature "Cambrian explosions" of rapid evolution.

We also found, not surprisingly, that the evolution was history dependent. In different runs of the experiment the same simple technologies would emerge, but in a different sequence. Because more complicated technologies are constructed from simpler ones, they would often be put together from different building blocks. (If bronze appears before iron in the real world, many artifacts are made of bronze; if iron appears before bronze, the same artifacts would be made of iron.) We also found that some complex needs for circuits such as adders or comparators with many inputs—different ones each time—would not be fulfilled at all.

And we found avalanches of destruction. Superior technologies replaced previously functioning ones. And this meant that circuits used only for these now obsolete technologies were themselves no longer needed, and so these in turn were replaced. This yielded avalanches we could study and measure.

In these ways we were able to examine the evolution of technology in action, and it bore out the story I gave earlier in this chapter.

A Different Form of Evolution?

I want to make some comments about the form of evolution I have been describing in this chapter. The first is that there is nothing predetermined about the exact sequence in which it unfolds. We cannot tell in advance which phenomena will be discovered and converted into the basis for new technologies. Nor in the vastness of combination possibilities can we tell which combinations will be created. Nor can we tell what opportunities these will open when they are realized. As a result of these indeterminacies, the evolution or unfolding of the collective is historically contingent. It depends upon small events in history: who encountered whom, who borrowed what idea, what authority decreed what. These small differences do not average out over time. They become built in to the collective, and because the technologies that enter depend upon those that exist, they propagate further differences. If we were to "replay" history a second time, we might end up with some similar collection of phenomena captured and therefore with roughly similar technologies. But the sequence and timing of their appearance would be different. And as a consequence, economic and social history would be different.

This does not mean the evolution of technology is completely random. The pipeline of technologies coming in the next decade is reasonably predictable. And current technologies have future improvement paths that will be followed more or less predictably. But overall, just as the collection of biological species in the far future is not predictable from the current collection, the collective of technology in the economic future is not predictable. Not only can we not forecast which combinations will be made, we also cannot forecast which opportunity niches will be created. And because potential combinations grow exponentially, this indeterminacy increases as the collective develops. Where three thousand years ago we could say with confidence that the technologies used a hundred years hence would resemble those currently in place, now we can barely predict the shape of technology fifty years ahead.

Another comment is that this sort of evolution does not cause change uniformly in time. Some of the time such a system is quiescent, with opportunity niches fulfilled and others quietly awaiting suitable combinations, and with small innovations happening—a novel combination here, replacement of a combination there. At other times the system is boiling with change. A novel combination of considerable consequence can appear—the steam engine, say— and a burst of change is unleashed. New niches form, new combinations appear, and much rearranging occurs. Change begets spates of change, and between these, quiescence begets quiescence.

Repeatedly throughout this book I have said that the collective of technology "evolves." I mean this literally. Just as biological organisms have evolved and built out in numbers and complexity, so too have technologies. We therefore have a second example of evolution on a grand scale over a lengthy time. Not eons, I grant, but still the drawn-out epoch of human existence.

How exactly does technological evolution compare with biological evolution? Is it different?

Our mechanism—I have been calling it *combinatorial evolution*—is about things creating novel things by combinations of themselves. The pure equivalent in biology would be to select an organ that had proved particularly useful in lemurs, say, and another organ from iguanas, and another one from macaque monkeys, and use these in combination with others to create a new creature. This seems absurd, but biology does form novel structures by combination from time to time. Genes are swapped and recombined in certain primitive bacteria by a mechanism called horizontal gene transfer. And genetic regulatory networks (roughly the "programs" that determine the sequence in which genes are switched on in an organism) occasionally add parts by combination.

More familiarly, larger structures are created as combinations of simpler ones in biology. The eukaryotic cell appeared as a combination of simpler structures, and multicellular organisms appeared as combinations of single-celled ones. And within organisms, functional features—the combination of small bones that form the mechanism that connects the eardrum to the auditory nerves, for example—can form from individual parts lying at hand. "In our universe," says the molecular biologist François Jacob, "matter is arranged in a hierarchy of structures by successive integrations." This is correct. Combinatorial evolution is by no means absent in biology.

Still, the creation of these larger combined structures is rarer in biological evolution—much rarer—than in technological evolution. It happens not on a daily basis but at intervals measured in millions of years. This is because combination in biology must work through a Darwinian bottleneck. Combination (at least for the higher organisms) cannot select pieces from different systems and combine these at one go. It is hemmed in by the strictures of genetic evolution: a new combination must be created by incremental steps; these steps must produce something viable—some living creature—at all stages; and new structures must be elaborated in steps out of components that exist already. In biology, combinations do form, but not routinely and by no means often, and not by the direct mechanisms we see in technology. Variation and selection are foremost, with combination happening at very occasional intervals but often with spectacular results.

In technology, by contrast, combinations are the norm. Every novel technology and novel solution is a combination, and every capturing of a phenomenon uses a combination. In technology, combinatorial evolution is foremost, and routine. Darwinian variation and selection are by no means absent, but they follow behind, working on structures already formed.

This account of the self-creation of technology should give us a different feeling about technology. We begin to get a feeling of ancestry, of a vast body of things that give rise to things, of things that add to the collection and disappear from it. The process by which this happens is neither uniform nor smooth; it shows bursts of accretion and avalanches of replacement. It continually explores into the unknown, continually uncovers novel phenomena, continually creates novelty. And it is organic: the new layers form on top of the old, and creations and replacements overlap in time. In its collective sense, technology is not merely a catalog of individual parts. It is a metabolic chemistry, an almost limitless collective of entities that interact to produce new entities—and further needs. And we should not forget that needs drive the evolution of technology every bit as much as the possibilities for fresh combination and the unearthing of phenomena. Without the presence of unmet needs, nothing novel would appear in technology.

One last thought. I said that technology is self-creating. It is a living webwork that weaves itself out of itself. Can we say from this that in some sense—some literal sense—technology is alive?

There is no formal definition of "life." Rather, we judge whether something is "alive" by asking whether it meets certain criteria. In systems language we can ask: is the entity self-organizing (are there simple rules by which it puts itself together?), and autopoietic (is it self-producing?). Technology (the collective) passes both these tests. In more mundane language we can ask: does the entity reproduce, grow, respond and adapt to its environment, and take in and give out energy to maintain its being? Technology (again the collective) also passes these tests. It reproduces itself in the sense that its individual elements, like cells in an organism, die and are replaced. Its elements grow—they grow without cease. It adapts closely to its environment, both collectively and in its individual elements. And it exchanges energy with its environment: the energy it takes to form and run each technology, and the physical energy each technology gives back.

By these criteria technology is indeed a living organism. But it is living only in the sense that a coral reef is living. At least at this stage of its development—and I for one am thankful for this—it still requires human agency for its buildout and reproduction.

10

THE ECONOMY EVOLVING AS ITS
TECHNOLOGIES EVOLVE

We have been looking very directly at the evolution of technology in the last chapter. There is an alternative way to perceive this evolution, and that is through the eyes of the economy. An
economy mirrors the changes in its technologies, the additions and
replacements I have been talking about. And it does this not merely
by smoothly readjusting its patterns of production and consumption
or by creating fresh combinations, as we saw in Chapter 8. It alters its
structure—alters the way it is arranged—as its technologies evolve,
and does this at all times and at all levels.

So I want to go back to the steps in the evolution of technology
and look at how they play out in the economy. What we will see is a
natural process of structural change in the economy driven by the
evolutionary processes we have just delved into. To explore this we
will first need to think of the economy in a way that differs from the
standard one.

The Economy as an Expression of Its Technologies

The standard way to define the economy—whether in dictionaries
or economics textbooks—is as a "system of production and distribution and consumption" of goods and services. And we picture this

system, "the economy," as something that exists in itself as a backdrop to the events and adjustments that occur within it. Seen this way, the economy becomes something like a gigantic container for its technologies, a huge machine with many modules or parts that are its technologies—its means of production. When a new technology (the railroad for transportation, say) comes along, it offers a new module, a new upgrade, for a particular industry: the old specialized module it replaces (canals) is taken out and the new upgrade module is slid in. The rest of the machine automatically rebalances and its tensions and flows (prices, and goods produced and consumed) readjust accordingly.

This view is not quite wrong. Certainly it is the way I was taught to think about the economy in graduate school, and it is very much the way economics textbooks picture the economy today. But it is not quite right either. To explore structural change, I want to look at the economy in a different way.

I will define the economy as *the set of arrangements and activities by which a society satisfies its needs.* (This makes economics the study of this.) Just what are these arrangements? Well, we could start with the Victorian economists' "means of production," the industrial production processes at the core of the economy. Indeed, my definition would not have surprised Karl Marx. Marx saw the economy as issuing from its "instruments of production," which would have included the large mills and textile machinery of his day.

But I want to go beyond Marx's mills and machinery. The set of arrangements that form the economy include all the myriad devices and methods and all the purposed systems we call technologies. They include hospitals and surgical procedures. And markets and pricing systems. And trading arrangements, distribution systems, organizations, and businesses. And financial systems, banks, regulatory systems, and legal systems. All these are arrangements by which we fulfill our needs, all are means to fulfill human purposes. All are therefore by my earlier criterion "technologies," or purposed systems. I talked about this in Chapter 3, so the idea should not be too unfamiliar. It means that the New York Stock Exchange and the specialized provisions of contract law are every bit as much means to human purposes as are steel mills and textile machinery. They too are in a wide sense technologies.

If we include all these "arrangements" in the collective of technology, we begin to see the economy not as container for its technologies, but as something constructed from its technologies. The economy is a set of activities and behaviors and flows of goods and services mediated by—draped over—its technologies. It follows that the methods, processes, and organizational forms I have been talking about *form* the economy.

The economy is an expression of its technologies.

I am not saying that an economy is identical to its technologies. There is more to an economy than this. Strategizing in business, investing, bidding, and trading—these are all activities and not purposed systems. What I *am* saying is that the structure of the economy is formed by its technologies, that technologies, if you like, form the economy's skeletal structure. The rest of the economy—the activities of commerce, the strategies and decisions of the various players in the game, the flows of goods and services and investments that result from these—form the muscle and neural structure and blood of the body-economic. But these parts surround and are shaped by the set of technologies, the purposed systems, that form the structure of the economy.

The shift in thinking I am putting forward here is not large; it is subtle. It is like seeing the mind not as a container for its concepts and habitual thought processes but as something that emerges from these. Or seeing an ecology not as containing a collection of biological species, but as forming from its collection of species. So it is with the economy. The economy forms an ecology for its technologies, it forms out of them, and this means it does not exist separately. And as with an ecology, it forms opportunity niches for novel technologies and fills these as novel technologies arise.

This way of thinking carries consequences. It means that the economy emerges—wells up—from its technologies. It means that the economy does more than readjust as its technologies change, it continually forms and re-forms as its technologies change. And it means that the character of the economy—its form and structure—change as its technologies change.

In sum, we can say this: As the collective of technology builds, it creates a structure within which decisions and activities and flows of goods and services take place. It creates something we call "the economy." The economy in this way emerges from its technologies. It constantly creates itself out of its technologies and decides which new technologies will enter it. Notice the circular causality at work here. Technology creates the structure of the economy, and the economy mediates the creation of novel technology (and therefore its own creation). Normally we do not see this technology-creating-the-economy-creating-technology. In the short term of a year or so the economy appears given and fixed; it appears to be a container for its activities. Only when we observe over decades do we see the arrangements and processes that form the economy coming into being, interacting, and collapsing back again. Only in the longer reaches of time do we see this continual creation and re-creation of the economy.

Structural Change

What happens to this system that forms from its technologies as new technologies enter? We would still see the same adjustments and fresh combinations I spoke about in Chapter 8 of course; they are perfectly valid. But we would see something more: the addition of new technologies setting in motion a train of changes to the structure of the economy, to the set of arrangements around which the economy forms.

We are coming into territory here—that of structural change—that economic theory does not usually enter. But this is not empty territory. It is inhabited by historians, in our case by economic historians. Historians see the introduction of new technologies not simply as causing readjustments and growth, but as causing changes in the composition of the economy itself—in its structure. But they proceed for the most part on an ad hoc, case-by-case basis. Our way of thinking about the economy and technology by contrast gives us a means to think abstractly about structural change.

In practice a new technology may call forth new industries; it may require that new organizational arrangements be set up; it may cause new technical and social problems and hence create new opportunity niches; and all these themselves may call forth further compositional changes. We can capture this sequence of changes if we borrow the steps in the evolution of technology from the previous chapter, but now see them through the eyes of the economy. So let us do this.

Suppose now once again that a new technology enters the economy. It replaces old technologies that carried out the same purpose (and possibly renders obsolete the industries and other technologies that depended upon these). And of course, the new technologies cause the sort of economic readjustments I spoke about earlier. (So far, these are steps 1, 2, 4, and 6, from earlier.)

But also, as in steps 5 and 3:

- The new technology provides potential new elements that can be used internally in other technologies. It thereby acts to call forth other technologies that use and accommodate it. In particular it may give rise to novel organizations that contain it.

- The new technology may set up opportunity niches for further technologies. Such opportunities arise in various ways, but in particular they may arise from the new technology causing novel technical, economic, or social problems. And so, the new technology sets up needs for further technologies to resolve these.

I have really just restated in economic terms the steps that form the mechanism by which the collective of technology evolves. But now the emphasis is different because I am reinterpreting these steps to describe how a new structure for the economy forms. When a novel technology enters the economy, it calls forth novel arrangements—novel technologies and organizational forms. The new technology or new arrangements in turn may cause new problems. These in their turn are answered by further novel arrangements (or by existing technologies modified for the purpose), which in their turn may open need for yet more novel technologies. The whole moves forward in a sequence of problem and solution—of challenge and response—and it is this sequence we call structural change. In this way the economy forms and re-forms itself in spates of change, as novelty, new arrangements to accommodate this, and the opening of opportunity niches follow from each other.

Let me make this concrete with a particular example. When workable textile machinery began to arrive around the 1760s in Britain, it offered a substitute for the cottage-based methods of the time, where wool and cotton were spun and woven at home by hand in the putting-out system. But the new machinery at first was only partly successful; it required a larger scale of organization than did cottage hand work. And so it presented an opportunity for—and became a component in—a higher-level organizational arrangement, the textile factory or mill. The factory itself as a means of organization—a technology—in turn required a means to complement its machinery: it called for factory labor. Labor of course already existed in the economy, but it did not exist in sufficient numbers to supply the new factory system. The necessary numbers were largely drawn from agriculture, and this in turn required accommodation near the mills. Worker dormitories and worker housing were therefore provided, and from the combination of mills, workers, and their housing, industrial cities began to grow. A new set of societal means of organization had appeared—a new set of arrangements—and with these the structure of the Victorian industrial economy began to emerge. In this way the character of an era—a set of arrangements compatible with the superior technology of industrial machinery—fell into place.

But an era is never finished. Manufacturing laborers, many of them children, worked often in Dickensian conditions. This presented a strongly felt need for reform, not only of "the moral conditions of the lower classes" but of their safety as well. In due course the legal system responded with further arrangements, labor laws designed to prevent the worst of the excesses. And the new working class began to demand a larger share of the wealth the factories had created. They made use of a means by which they could better their conditions: trade unions. Labor was much easier to organize in factories than in isolated cottages, and over the course of decades it became a political force.

In this way the original arrival of textile machinery not only replaced cottage hand manufacturing, it set up an opportunity for a higher-level set of arrangements—the factory system—in which the machinery became merely a component. The new factory system in turn set up a chain of needs—for labor and housing—whose solutions created further needs, and all this in time became the Victorian industrial system. The process took a hundred years or more to reach anything like completion.

The reader might object that this makes structural change appear too simplistic—too mechanical. Technology A sets up a need for arrangements B; technology C fulfills this, but sets up further needs D and E; these are resolved by technologies F and G. Certainly such sequences do form the basis of structural change, but there is nothing simple about them. If we invoke recursion these new arrangements and technologies themselves call for subtechnologies and subarrangements. The factory system itself needed means of powering the new machinery, systems of ropes and pulleys for transmitting this power, means of acquiring and keeping track of materials, means of bookkeeping, means of management, means of delivery of the product. And these in turn were built from other components, and had their own needs. Structural change is fractal, it branches out at lower levels, just as an embryonic arterial system branches out as it develops into smaller arteries and capillaries.

And some of the responses are not economic at all. The very idea that hand craft could be mechanized spread from textiles to other industries and led to new machinery there. And again psychologically, factories created not just a new organizational set of arrangements but called for a new kind of person. Factory discipline, says historian David Lándes, "required and eventually created a new breed of worker. . . . No longer could the spinner turn her wheel and the weaver throw his shuttle at home, free of supervision, both in their own good time. Now the work had to be done in a factory, at a pace set by tireless, inanimate equipment, as part of a large team that had to begin, pause, and stop in unison—all under the close eye of overseers, enforcing assiduity by moral, pecuniary, occasionally even physical means of compulsion. The factory was a new kind of prison; the clock a new kind of jailer." The new technology caused more than economic change, it caused psychological change.

In talking about structural change then, we need to acknowledge that the set of changes may not all be tangible and may not all be "arrangements." And we need to keep in mind that changes may have multiple causes and a high multiplicity of effects. Nevertheless, I want to emphasize that we can think of the process of structural change in logical terms—theoretically if you like—using the steps laid out for the evolution of technology. Structural change in the economy is not just the addition of a novel technology and replacement of the old, and the economic adjustments that follow these. It is a chain of consequences where the arrangements that form the skeletal structure of the economy continually call forth new arrangements.

There is of course nothing inevitable—nothing predetermined— about the arrangements that fall into place and define the structure of the economy. We saw earlier that very many different combinations, very many arrangements, can solve the problems posed by technology. Which ones are chosen is in part a matter of historical small events: the order in which problems happen to be tackled, the predilections and actions of individual personalities. The actions, in other words, of chance. Technology determines the structure of the economy and thereby much of the world that emerges from this, but which technologies fall into place is not determined in advance.

Problems as the Answer to Solutions

I have talked about this unfolding of structure as a constant remaking of the arrangements that form the economy; one set of arrangements sets up conditions for the arrival of the next. There is no reason once set in motion this remaking should come to an end. The consequences of even one novel technology—think of the computer or the steam engine—can persist without letup.

It follows in turn that the economy is never quite at stasis. At any time its structure may be in some high degree of mutual compatibility, and hence close to unchanging. But within this stasis lie the seeds of its own disruption, as Schumpeter pointed out a hundred years ago. The cause is the creation of novel combinations—novel arrangements—or for Schumpeter the new "goods, the new methods of production or transportation, the new markets, the new forms of industrial organization" that set up a process of "industrial mutation" that "incessantly revolutionizes the economic structure *from within*, incessantly destroying the old one, incessantly creating a new one."

From within, the system is always poised for change.

But the argument I am giving implies more—much more—than Schumpeter said. The coming of novel technologies does not just disrupt the status quo by finding new combinations that are better versions of the goods and methods we use. It sets up a train of technological accommodations and of new problems, and in so doing it creates new opportunity niches that call forth fresh combinations which in turn introduce yet further technologies—and further problems.

The economy therefore exists always in a perpetual openness of change—in perpetual novelty. It exists perpetually in a process of self-creation. It is always unsatisfied. We can add to this that novel technologies of all degrees of significance enter the economy at any time alongside each other. The result is not just Schumpeter's disturbance of equilibrium, but a constant roiling of simultaneous changes, all overlapping and interacting and triggering further change. The result is change begetting change.

Curiously, we may not be very conscious of this constant roiling at any time. This is because the process of structural change plays out over decades, not months. It is more like the slow geological upheavals that take place under our feet. In the short term the structure in place has a high degree of continuity; it is a loosely compatible set of systems within which plans can be made and activities can take place. But at all times this structure is being altered. The economy is perpetually constructing itself.

Could this process of constant evolution of technology and remaking of the economy ever come to a halt? In principle, it could. But only in principle. This could only happen if no novel phenomena in the future were to be uncovered; or if the possibilities for further combinations were somehow exhausted. Or if our practical human needs were somehow fulfilled by the available technologies we possessed. But each of these possibilities is unlikely. Ever-open needs and the likely discovery of new phenomena will be sufficient to drive technology forward in perpetuity, and the economy with it.

Coming to a halt is unlikely for another reason. I have been stressing that every solution in the form of a new technology creates some new challenge, some new problem. Stated as a general rule, *every technology contains the seeds of a problem*, often several. This is not a "law" of technology or of the economy, much less one of the universe. It is simply a broad-based empirical observation—a regrettable one—drawn from human history. The use of carbon-based fuel technologies has brought global warming. The use of atomic power, an environmentally clean source of power, has brought the problem of disposal of atomic waste. The use of air transport has brought the potential of rapid worldwide spread of infections. In the economy, solutions lead to problems, and problems to further solutions, and this dance between solution and problem is unlikely to change at any time in the future. If we are lucky we experience a net benefit that we call progress. Whether or not progress exists, this dance condemns technology—and the economy as a result—to continuous change.

What I have been talking about in this chapter is really the evolution of technology seen through the eyes of the economy. Because the economy is an expression of its technologies, it is a set of arrangements that forms from the processes, organizations, devices, and institutional provisions that comprise the evolving collective; and it evolves as its technologies do. And because economy arises out of its technologies, it inherits from them self-creation, perpetual openness, and perpetual novelty. The economy therefore arises ultimately out of the phenomena that create technology; it is nature organized to serve our needs.

There is nothing simple about this economy. Arrangements are built one upon another: the commercial parts of the legal system are constructed on the assumption that markets and contracts exist; and markets and contracts assume that banking and investment mechanisms exist. The economy therefore is not a homogeneous thing. It is a structure—a magnificent structure—of interacting, mutually supporting arrangements, existing at many levels, that has grown itself from itself over centuries. It is almost a living thing, or at least an evolving thing, that changes its structure continually as its arrangements create further possibilities and problems that call forth further responses—yet further arrangements.

This evolution of structure is a constant remaking of the arrangements that form the economy, as one set of arrangements sets up the conditions for the arrival of the next. This is not the same as readjustment within given arrangements or given industries, and it is not the same as economic growth. It is continual, fractal, and inexorable. And it brings unceasing change.

Is there anything constant about structural change? Well, the economy forms its patterns always from the same elements—the predilections of human behavior, the basic realities of accounting, and the truism that goods bought must equal goods sold. These underlying base "laws" always stay the same. But the means by which they are expressed change over time, and the patterns they form change and re-form over time. Each new pattern, each new set of arrangements, then, yields a new structure for the economy and the old one passes, but the underlying components that form it—the base laws—remain always the same.

Economics as a discipline is often criticized because, unlike the "hard sciences" of physics or chemistry, it cannot be pinned down to an unchanging set of descriptions over time. But this is not a failing, it is proper and natural. The economy is not a simple system; it is an evolving, complex one, and the structures it forms change constantly over time. This means our interpretations of the economy must change constantly over time. I sometimes think of the economy as a World War I battlefield at night. It is dark, and not much can be seen over the parapets. From a half mile or so away, across in enemy territory, rumblings are heard and a sense develops that emplacements are shifting and troops are being redeployed. But the best guesses of the new configuration are extrapolations of the old. Then someone puts up a flare and it illuminates a whole pattern of emplacements and disposals and troops and trenches in the observers' minds, and all goes dark again. So it is with the economy. The great flares in economics are those of theorists like Smith or Ricardo or Marx or Keynes. Or indeed Schumpeter himself. They light for a time, but the rumblings and redeployments continue in the dark. We can indeed observe the economy, but our language for it, our labels for it, and our understanding of it are all frozen by the great flares that have lit up the scene, and in particular by the last great set of flares.

11

___

WHERE DO WE STAND WITH THIS

CREATION OF OURS?

I said at the outset the purpose of this book was to create a theory of technology—"a coherent group of general propositions"—that gives us a framework for understanding what technology is and how it works in the world. In particular the purpose was to create a theory of evolution valid for technology, and not borrowed from outside. The result has been, to use Darwin's phrase, "one long argument." At the cost of some repetition, let me give a brief summary of this argument here.

Theories start with general propositions or principles, and we started with three: that all technologies are combinations of elements; that these elements themselves are technologies; and that all technologies use phenomena to some purpose. This third principle in particular told us that in its essence, technology is a programming of nature. It is a capturing of phenomena and a harnessing of these to human purposes. An individual technology "programs" many phenomena; it orchestrates these to achieve a particular purpose.

Once new technologies, individual ones, exist they become potential building blocks for the construction of further new technologies. The result is a form of evolution, combinatorial evolution, whose base mechanism differs from the standard Darwinian one.

Novel technologies are created out of building blocks that are themselves technologies, and become potential building blocks for the construction of further new technologies. Feeding this evolution is the progressive capturing and harnessing of novel phenomena, but this requires existing technologies both for the capturing and the harnessing. From these last two statements we can say that technology creates itself out of itself. In this way the collection of mechanical arts that are available to a culture bootstraps itself upward from few building-block elements to many, and from simple elements to more complicated ones.

This form of evolution appears simple. But as with Darwinian evolution it is not; there are many details and mechanisms. The central mechanism is of course the one by which radically novel technologies originate. New "species" in technology arise by linking some need with some effect (or effects) that can fulfill it. This linking is a process, a lengthy one of envisioning a concept—the idea of a set of effects in action—and finding a combination of components and assemblies that will make the concept possible. The process is recursive. Getting a concept to work brings up problems, and the potential solution of these brings up further subproblems. The process goes back and forth between problems and solutions at different levels before it is complete.

Combination, putting together suitable parts and functionalities mentally or physically to form a solution, is at the heart of this. But it is not the only force driving technology's evolution. The other one is need, the demand for novel ways of doing things. And needs themselves derive more from technology itself than directly from human wants; they derive in the main from limitations encountered and problems engendered by technologies themselves. These must be solved by still further technologies, so that with technology need follows solution as much as solution follows need. Combinatorial evolution is every bit as much about the buildout of needs as about solutions to these.

The overall process by which all this happens is neither uniform nor smooth. At all times, the collective of technology is evolving by adding and dropping technologies, by creating opportunity niches for further technologies, and by uncovering novel phenomena. Bodies of technology are evolving too, in the narrower sense of continual development: they emerge, constantly change the "vocabularies" they provide, and are absorbed into the economy's industries. And individual technologies are evolving—developing—too. To deliver better performance, they continually change their internal parts and add more complex assemblies.

The result is a constant roiling at all levels. At all levels new combinations appear, new technologies are added, and old ones disappear. In this way technology constantly explores into the unknown, constantly creates further solutions and further needs, and along with this, perpetual novelty. The process is organic: the new layers form on top of the old, and creations and replacements overlap in time. In its collective sense, technology is not merely a catalog of individual parts. It is a metabolic chemistry, an almost limitless collective of entities that interact and build from what is there to produce new entities—and further needs.

The economy directs and mediates all this. It signals needs, tests ideas for commercial viability, and provides demands for new versions of technologies. But it is not a simple receptor of technology, not a machine that receives upgrades to its parts every so often. The economy is an expression of its technologies. Its skeletal structure consists in a mutually supporting set of arrangements—businesses, means of production, institutions, and organizations—that are themselves technologies in the broad sense. Around these the activities and actions of commerce take place. These "arrangements" create opportunities for further "arrangements," and the sequence by which they follow one another constitutes structural change in the economy. The resulting economy inherits all the qualities of its technologies. It too, on a long-term scale, seethes with change. And like technology, it is open, history-dependent, hierarchical, indeterminate. And ever changing.

Technology Becoming Biology—and Vice Versa

There is a possible objection to my argument. Many of the examples I have given in this book are historical—drawn from twentieth- and nineteenth-century technology. So as we move into the future, the vision I am putting forth of technologies as a chemistry of functionalities, programmable in different configurations for different purposes, will possibly no longer be valid.

Actually, the opposite is true. The vision is becoming ever more appropriate as technology progresses. Digitization allows functionalities to be combined even if they come from different domains, because once they enter the digital domain they become objects of the same type—data strings—that can therefore be acted upon in the same way. Telecommunications allows these digital elements to be combined remotely so that virtually any executable anywhere can trigger another. And with sensing devices, systems can now perceive their environment, albeit primitively, and configure their actions. The result is a hitching together of functionalities from different domains and from widely separated locations into temporary networks, connected collections of things-in-conversation-with-things that sense their environment and react appropriately. Thus modern passenger aircraft navigation is a set of functional elements—an onboard gyro system, GPS system, several navigational satellites and ground stations, atomic clocks, autopilot and fly-by-wire system, actuators that position the control surfaces—all in "conversation" with each other, querying each other, triggering each other, executing each other, much as a set of subroutines in a computer algorithm query and trigger and execute each other.

The representative technology is no longer a machine with fixed architecture carrying out a fixed function. It is a system, a network of functionalities—a metabolism of things-executing-things—that can sense its environment and reconfigure its actions to execute appropriately.

Such networks of things-executing-things can be designed from the top down to sense their environment and react appropriately. But this is difficult. When a network consists of thousands of separate interacting parts and the environment changes rapidly, it becomes almost impossible to design top-down in any reliable way. Therefore, increasingly, networks are being designed to "learn" from experience which simple interactive rules of configuration operate best within different environments. Equipped with such rules, they can react appropriately to what is sensed. Does this constitute some form of "intelligence"? To some degree it does. One simple definition of bio-logical cognition—as say with *E. coli* bacteria sensing an increased concentration of glucose and moving toward that—is being able to sense an environment and react appropriately. Thus, as modern technology organizes itself increasingly into networks of parts that sense, configure, and execute appropriately, it displays some degree of cognition. We are moving toward "smart" systems. The arrival of genomics and nanotechnology will enhance this. In fact, not only will these systems in the future be self-configuring, self-optimizing, and cognitive, they will be self-assembling, self-healing, and self-protecting.

My purpose here is not to point to some science fiction future, or to discuss the implications of these trends. Others have done this else-where. I want to call attention to something else: words such as self-configuring, self-healing, and cognitive are not ones we would have associated with technology in the past. These are biological words. And they are telling us that as technology becomes more sophisti-cated, it is becoming more biological. This seems paradoxical. Surely technology by its very essence is mechanistic; therefore surely as it becomes more complicated it becomes simply more complicatedly mechanistic. So how can technologies be becoming more biological?

There are two answers. One is that all technologies are in a sense simultaneously mechanistic and organic. If you examine a technol-ogy from the top down, you see it as an arrangement of connected parts interacting and intermeshing with each other to some purpose. In this sense it becomes a clockwork device—it becomes mechanistic. If you examine it mentally from the bottom up, however, from how these parts are put together, you see these as integral parts—integral organs—forming a higher, functioning, purposed whole. It becomes a functioning body—it becomes organic. Whether a technology is mechanistic or organic therefore depends on your point of view. The other answer is purely biological: technologies are acquiring properties we associate with living organisms. As they sense and react to their environment, as they become self-assembling, self-configuring, self-healing, and "cognitive," they more and more resemble living organisms. The more sophisticated and "high-tech" technologies become, the more they become biological. We are beginning to appreciate that technology is as much metabolism as mechanism.

There is a symmetric side to this. As biology is better understood, we are steadily seeing it as more mechanistic. Of course, the idea that biological organisms consist of connected parts that interact like those of a machine is not new. It goes back at least to the 1620s, to the time of Mersenne and Descartes when philosophers were starting to think of living things as possible machines. What *is* new is that we now understand the working details of much of the machinery. Since the 1950s, we have teased out piece by piece the finer workings of DNA and protein manufacture within the cell, some of the elaborate controls for gene expression, and the functions of the parts of the brain. This work is far from complete, but it reveals organisms and organelles as highly elaborate technologies. In fact, living things give us a glimpse of how far technology has yet to go. No engineering technology is remotely as complicated in its workings as the cell.

Conceptually at least, biology is becoming technology. And physically, technology is becoming biology. The two are starting to close on each other, and indeed as we move deeper into genomics and nanotechnology, more than this, they are starting to intermingle.

The Generative Economy

If technology is changing to become more configurable and more biological, does the economy in any way reflect this? It should, if indeed it is an expression of its technologies. And indeed it does.

Technology has long shifted away from the Victorian era's dominance of bulk material processing, and now it is shifting again: from single-purpose fixed processes or machines into raw functionalities that can be programmed to different purposes in different combinations. Reflecting this, the economy—the high-tech part of it, at least—is more about the putting together of things than about the refining of fixed operations. Business operations—commercial banks, oil companies, insurance companies—of course still reflect the era of large, fixed technologies. But increasingly, as with the operations of putting together a startup company, or venture capital, or financial derivatives, or digitization, or combinatorial biology, they are about combining functionalities—actions and business processes—for short-term reconfigurable purposes.

The economy, in a word, is becoming generative. Its focus is shifting from optimizing fixed operations into creating new combinations, new configurable offerings.

For the entrepreneur creating these new combinations in a startup company, little is clear. He often does not know who his competitors will be. He does not know how well the new technology will work, or how it will be received. He does not know what government regulations will apply. It is as if he is placing bets in a casino game where the rules and payoffs are not clear until after the bets have been laid. The environment that surrounds the launching of a new combinatorial business is not merely uncertain; particular aspects of it are simply unknown.

This means that the decision "problems" of the high-tech economy are not well defined. As such (perhaps shockingly to the reader), they have no optimal "solution." In this situation the challenge of management is not to rationally solve problems but to make sense of an undefined situation—to "cognize" it, or frame it into a situation that can be dealt with—and to position its offerings accordingly. Again here is a seeming paradox. The more high-tech technology becomes, the less purely rational becomes the business of dealing with it. Entrepreneurship in advanced technology is not merely a matter of decision making. It is a matter of imposing a cognitive order on situations that are repeatedly ill-defined. Technology thinker John Seely Brown tells us that "management has shifted from making product to making sense."

In the generative economy, management derives its competitive advantage not from its stock of resources and its ability to transform these into finished goods, but from its ability to translate its stock of deep expertise into ever new strategic combinations. Reflecting this, national wealth derives not so much from the ownership of resources as from the ownership of specialized scientific and technical expertise. Companies too draw their competitive advantage from their ownership of technical expertise. Often in putting together a new combination they lack specific expertises, and in the pressure of competition they do not have time to develop them in-house. So they buy small companies, or form strategic alliances with other companies that do possess the requisite craft. Such alliances are often put together for a specific purpose, then reconfigured or dropped. And so we see combination at the firm level, expressed as a continuously reconfiguring series of loose alliances—ephemeral, and occasionally highly successful.

And so the nature of modern technology is bringing a new set of shifts: In the management of businesses, from optimizing production processes to creating new combinations—new products, new functionalities. From rationality to sense-making; from commodity-based companies to skill-based companies; from the purchase of components to the formation of alliances; from steady-state operations to constant adaptation. None of these shifts is abrupt, and in fact the elements of the new and old styles exist together in the economy: the two worlds of business overlap and are heavily interrelated. But as a more technological economy comes to the fore, we are shifting from the machine-like economy of the twentieth century with its factory nodes and input-output linkages to an organic, interrelated economy of the twenty-first century. Where the old economy was a machine, the new one is a chemistry, always creating itself in new combinations, always discovering, always in process.

Economics itself is beginning to respond to these changes and reflect that the object it studies is not a system at equilibrium, but an evolving, complex system whose elements—consumers, investors, firms, governing authorities—react to the patterns these elements create. Its standard doctrines were built upon the bedrock principles of predictability, order, equilibrium, and the exercise of rationality; and this suited an economy that consisted of bulk-process technologies that remained much the same from year to year. But as the economy becomes more combinatorial and technology more open, new principles are entering the foundations of economics. Order, closedness, and equilibrium as ways of organizing explanations are giving way to open-endedness, indeterminacy, and the emergence of perpetual novelty.

Pure Order Versus Messy Vitality

Not only is our understanding of the economy changing to reflect a more open, organic view. Our interpretation of the world is also becoming more open and organic; and again technology has a part in this shift. In the time of Descartes we began to interpret the world in terms of the perceived qualities of technology: its mechanical linkages, its formal order, its motive power, its simple geometry, its clean surfaces, its beautiful clockwork exactness. These qualities have projected themselves on culture and thought as ideals to be used for explanation and emulation—a tendency that was greatly boosted by the discoveries of simple order and clockwork exactness in Galilean and Newtonian science. These gave us a view of the world as consisting of parts, as rational, as governed by Reason (capitalized in the eighteenth century, and feminine) and by simplicity. They engendered, to borrow a phrase from architect Robert Venturi, prim dreams of pure order.

And so the three centuries since Newton became a long period of fascination with technique, with machines, and with dreams of the pure order of things. The twentieth century saw the high expression of this as this mechanistic view began to dominate. In many academic areas—psychology and economics, for example—the mechanistic interpretation subjugated insightful thought to the fascination of technique. In philosophy, it brought hopes that rational philosophy could be founded on—constructed from—the elements of logic and later of language. In politics, it brought ideals of controlled, engineered societies; and with these the managed, controlled structures of socialism, communism, and various forms of fascism. In architecture, it brought the austere geometry and clean surfaces of Le Corbusier and the Bauhaus. But in time all these domains sprawled beyond any system built to contain them, and all through the twentieth century movements based on the mechanistic dreams of pure order broke down.

In its place is growing an appreciation that the world reflects more than the sum of its mechanisms. Mechanism, to be sure, is still central. But we are now aware that as mechanisms become interconnected and complicated, the worlds they reveal are complex. They are open, evolving, and yield emergent properties that are not predictable from their parts. The view we are moving to is no longer one of pure order. It is one of wholeness, an organic wholeness, and imperfection. Here is Venturi again, speaking of architecture.

> I like elements which are hybrid rather than "pure," compromising rather than "clean," distorted rather than "straightforward," ambiguous rather than "articulated," perverse as well as impersonal, boring as well as "interesting," conventional rather than "designed," accommodating rather than excluding, redundant

> rather than simple, vestigial as well as innovating, inconsistent and equivocal rather than direct and clear. I am for messy vitality over obvious unity. I include the non sequitur and proclaim the duality. I am for richness of meaning rather than clarity of meaning; for the implicit function as well as the explicit function.

Messy vitality, says Venturi; and richness of meaning. Yes. I too am wholeheartedly for these. We are replacing our image of perfection with an image of wholeness, and within that wholeness a messy vitality. This shift in thinking has more to do with the influence of evolutionary biology and the exhaustion of the simple mechanistic view than with any influence from modern technology. But it is reinforced nonetheless by the qualities of modern technology: its connectedness, its adaptiveness, its tendency to evolve, its organic quality. Its messy vitality.

Where Do We Stand with Technology?

We are changing then in how we view the world through technology. But what about technology itself? How do we see it? Where do we stand with this creation of ours?

We feel of course a deep ambivalence toward technology, one that steadily grows. But this ambivalence comes not from our relationship with technology, not directly at any rate; it comes from our relationship with nature. This should not be altogether surprising. If technology is nature organized for our purposes, to a very large degree our relationship to this use of nature should determine how we think of technology.

In 1955, Martin Heidegger gave a lecture entitled "The Question Concerning Technology." The essence of technology, he said, is by no means anything technological. It is a way of seeing nature, of letting all that is in nature reveal itself as a potential resource for us to use as human beings. This is regrettable. "Nature becomes a gigantic gasoline station," and we see it as an exploitable resource, as mere "standing reserve" to be used for our purposes. Technology, or *techne* in the Greek sense of craft or knowing in action, ceases to be a "bringing forth of the true into the beautiful," as it would be in the hands of an ancient silversmith making a sacrificial chalice. Instead of fitting itself to the world, technology seeks to fit the world to itself. Heidegger does not say the problem lies with technology. It lies with the attitude technology has brought with it. Where once we respected—indeed revered—nature, now we have "set upon" nature and reduced it to something that merely stands by for our use.

Worse, we have created a thing, technology, that responds not primarily to human need but to its own needs. "Technology is in no sense an instrument of man's making or in his control," says Heidegger's translator, paraphrasing him. "It is rather that phenomenon, ruled from out of Being itself, that is centrally determining all of Western history." Others—notably the French sociologist Jacques Ellul—have said much the same, in words no less dramatic. Technology is a Thing directing human life, a Thing to which human life must bow and adapt, "an 'organism' tending toward closure and self-determination; it is an end in itself."

Yet—and Heidegger concedes this—our relationship with this thing, technology, has also served us well. Technology has created our economy, and with it all our wealth and security. It has brought us lives much longer than those our ancestors lived, lives blessedly free of the miseries they faced.

These two views, that technology is a thing directing our lives, and simultaneously a thing blessedly serving our lives, are simultaneously valid. But together they cause an unease, an ongoing tension, that plays out in our attitudes to technology and in the politics that surround it.

This tension does not just come from technology causing us to exploit nature and from it determining much of our lives. It arises, as

I said in Chapter 1, because for all of human existence we have been at home in nature—we *trust* nature, not technology. And yet we look to technology to take care of our future—we *hope* in technology. So we hope in something we do not quite trust. There is an irony here. Technology, as I have said, is the programming of nature, the orchestration and use of nature's phenomena. So in its deepest essence it is natural, profoundly natural. But it does not *feel* natural.

If we merely used nature's phenomena in raw form, to power water wheels or propel sailing ships, we would feel more at home with technology, and our trust and hope would be less at odds. But now, with the coming of genetic engineering, machine intelligence, bionics, climate engineering, we are beginning to use technology—use nature—to intervene directly within nature. And to our primate species, at home in a habitat of trees and grasses and other animals, this feels profoundly unnatural. This disturbs our deep trust.

Unconsciously we react to this deep unease in many ways. We turn to tradition. We turn to environmentalism. We hearken to family values. We turn to fundamentalism. We protest. Behind these reactions, justified or not, lie fears. We fear that technology separates us from nature, destroys nature, destroys *our* nature. We fear this phenomenon of technology that is not in our control. We fear we are unleashing some thing of disembodied action somehow taking on a life of its own and coming somehow to control *us*. We fear technology as a living thing that will bring us death. Not the death of nothingness, but a worse death. The death that comes with no-freedom. The death of will.

We sense this unconsciously. And the popular myths of our time point to this. Whether in fiction or movies, if we examine the stories we tell ourselves, we see the question is not whether we should possess technology or not. It is whether we should accept technology as faceless and will-deadening versus possess technology as organic and life-enhancing. In the movie *Star Wars*, the malign aspect of technology is the Death Star. It is an object huge and disconnected from humanness that reduces its clients to clones—recognizably human, but all identically in thrall to the machine, all drained of color and drained of will. Its protagonist, Darth Vader, is not a full human being either. He is constructed—part technology, part human body. The heroes, Luke Skywalker and Han Solo, by contrast are fully human. They have individuality, they have will, and they hang with creatures in a haunt called the Mos Eisley Cantina—creatures that are strange, distorted, and perverse, but that brim with messy vitality. If you look at the heroes, they have technology as well. But their technology is different. It is not hidden and dehumanizing; their starships are rickety and organic and have to be kicked to get running. This is crucial. Their technology is human. It is an extension of their natures, fallible, human, individual, and therefore beneficent. They have not traded their humanness for technology, nor surrendered their will to technology. Technology has surrendered to *them*. And in doing so it extends their naturalness.

Thus our reaction to technology as represented unconsciously in popular myth does not reject technology. To have no technology is to be not-human; technology is a very large part of what makes us human. "The Buddha, the Godhead, resides quite as comfortably in the circuits of a digital computer or the gears of a cycle transmission as he does at the top of a mountain or in the petals of a flower," says Robert Pirsig. Technology is part of the deeper order of things. But our unconscious makes a distinction between technology as enslaving our nature versus technology as extending our nature. This is the correct distinction. We should not accept technology that deadens us; nor should we always equate what is possible with what is desirable. We are human beings and we need more than economic comfort. We need challenge, we need meaning, we need purpose, we need alignment with nature. Where technology separates us from these it brings a type of death. But where it enhances these, it affirms life. It affirms our humanness.

NOTES

Preface

2   *Technologies in other . . . combinations*: I discussed this idea with Stuart Kauffman in 1987. Kauffman has since followed up with further thoughts on the self-creating aspect of technology in several of his writings.

4   *beautiful case studies . . . historians*: See in particular Aitken, Constant, Hughes, Landes, Rhodes, and Tomayko.

Chapter 1: Questions

9   *monkey with tiny electrodes*: Meel Velliste, *et al.* "Cortical Control of a Prosthetic Arm for Self-feeding," *Nature*, 453, 1098–101, June 19, 2008.

13   *Georges Cuvier*: The quote is from his *Tableau Élémentaire de l'Histoire Naturelle des Animaux*, Baudouin, Paris, 1798.

15   *The word "evolution" . . . general meanings*: See Stephen Jay Gould, "Three Facets of Evolution," in *Science, Mind, and Cosmos*, J. Brockman and K. Matson, eds., Phoenix, London, 1995.

16   *Gilfillan traced the . . . ship*: Gilfillan, 1935a.

17   *abrupt radical novelty . . . evolutionists*: Basalla, whose 1988 *The Evolution of Technology* is the most complete theory to date, ends up having to confess (p. 210) "our inability to account for the emergence of novel artifacts."

19   *this idea, like . . . people*: An early example is Thurston, 1883, p. 3.

20   *Without outside disturbances . . . settle*: Schumpeter 1912. Schumpeter had recently visited the doyen of equilibrium economics, Léon Walras, in Switzerland, who had told him that "of course economic life is essentially passive and merely adapts itself to the natural and social influences which may be acting on it." See Richard Swedberg, *Schumpeter: A Biography*, University Press, Princeton, 1991, p. 32.

20   *Invention, said historian . . . Usher*: Usher, p. 11; see also Gilfillan, 1935b, p. 6; and McGee.

20   *William Fielding Ogburn*: Ogburn, p. 104.
23   *"a coherent group . . . propositions"*: This definition of "theory" is from *Dic tionary.com Unabridged* (v 1.1). Random House, Inc., accessed 2008.

Chapter 2: Combination and Structure

27   *What is technology?*: Technology is, according to the *American College Dictionary*, "that branch of knowledge that deals with the industrial arts"; or, per *Webster*, "the science of the application of knowledge to practical purposes: applied science"; or according to *Britannica*, "the systematic study of techniques for making and doing things." "The totality of the means" is from *Webster's Third New International Dictionary*, Merriam-Webster, 1986.

27   *can technology really be knowledge*: I disagree that technology is knowledge. Knowledge is necessary for a technology—knowledge of how to construct it, think about it, deal with it—but this does not make technology the same as knowledge. We could say that knowledge is also necessary for mathematics—knowledge of theorems, structures, and methods—but that does not make mathematics the same as knowledge. Knowledge is the possession of information, facts, understandings, and the holding of these about something is not the same as that something. For me, also, a technology is something that can be executed. If you jump from an airplane you want a parachute, not knowledge of how to make a parachute.

30   *The radio processes the signal*: Most modern radios also include a heterodyning stage that converts the radio-frequency signal to a fixed intermediate frequency that subsequent circuits are optimized to handle.

32   *we find the . . . useful*: Nietzsche comments: "Every concept originates through our equating what is unequal. No leaf ever wholly equals another, and the concept 'leaf' is formed through an arbitrary abstraction from the individual differences, through forgetting the distinctions; and now it gives rise to the idea that in nature there might be something besides the leaves which would be 'leaf'—some kind of original form after which all leaves have been woven, marked, copied, colored, curled, and painted, but by unskilled hands, so that no copy turned out to be a correct, reliable, and faithful image of the original form." From "On Truth and Lie in an Extra-Moral Sense," *The Portable Nietzsche*, Penguin, New York, 1976, p. 46.

36   *"chunking" in cognitive psychology*: The idea goes back to the 1950s; K. S. Lashley, "The problem of serial order in behavior," in L. A. Jeffress, ed., *Cerebral Mechanisms in Behavior*, Wiley, New York, 1951; also F. Gobet, *et al.*, "Chunking mechanisms in human learning," *Trends in Cognitive Sciences*, 5, 6: 236–243, 2001.

37   *Adam Smith*: Smith, *The Wealth of Nations*, 1776, Chapter 1.

37   *modules of technology . . . units*: Baldwin and Clark show that modularization is increasing over time.

38   *The hierarchy that forms*: Herbert Simon talked of hierarchical systems, but not of recursiveness.

38 *Recursiveness will be . . . principle*: The related property, that the component entities resemble in some sense the higher level entities, is called "self-similarity." Where I say the structure is fractal, I mean this loosely. Strictly speaking, a fractal is a geometric object.

Chapter 3: Phenomena

47 *Phenomena are the . . . source*: For a compilation of physical effects, see Joachim Schubert's *Dictionary of Effects and Phenomena in Physics*, Wiley, New York, 1987.

47 *the astronomers Geoffrey . . . Butler*: Interview with Geoffrey Marcy, February 20, 2008.

58 *In 1831 Faraday discovered*: James Hamilton, *Faraday: The Life*, HarperCollins, London, 2002.

60 *engineering professor John G. Truxal*: The quote is from Truxal, "Learning to Think Like an Engineer: Why, What, and How?" *Change* 3, 2:10–19, 1986.

62 *It took Millikan five years*: Robert P. Crease, *The Prism and the Pendulum*, Random House, New York, 2003.

64 *science is . . . technology*: "Forced at the point of a gun to choose between two crude misconceptions," says technology philosopher Robert McGinn (p. 27), "one would have to opt for the heretical notion that 'science is applied technology' over the conventional wisdom that 'technology is applied science.'"

65 *Mokyr has pointed . . . technologies*: Mokyr, 2004.

Chapter 4: Domains, or Worlds Entered for What Can Be Accomplished There

70 *A domain will . . . cluster*: Of course, domains suffer the usual problems of collective nouns. (Who exactly is a conservative? What exactly constitutes renaissance architecture?) Domains may also overlap. Roller bearings belong to several commonly used domains.

72 *Fly-by-wire allowed aircraft control systems*: Tomayko.

74 *In 1821 Charles Babbage*: Joel Shurkin, *Engines of the Mind,* Norton, New York, 1996, p. 42; Doron Swade, *The Difference Engine*, Penguin Books, New York, 2002, p. 10.

74 *Jules Verne's*: Dover's 1962 edition of Verne's *From the Earth to the Moon* carries many of the original 1860s French illustrations.

77 *"the grammar of painting"*: Henry James, "The Art of Fiction," *Longman's Magazine* 4, September 1884.

77 *the biochemist Erwin Chargaff*: The quote is from his "Preface to a Grammar of Biology," *Science* 172, May 14, 1971.

77 *Where do such . . . from?*: Sometimes grammars exist with no obvious backing of nature. The grammar of a programming language such as C++ is artificial, and based on an agreed-upon set of principles. See Stroustrup.

Grammars do not necessarily derive from our "official" understanding of how nature works, from science. Most of the newer grammars—of nano-technology, say, or of optical data transmission—do. But the principles that govern older technologies, such as the smelting of metals or leather tanning, were derived mainly from practice, from the casual observation of nature.

77 *But this understanding . . . theory*: The aircraft rule is from Vincenti, p. 218.

78 *"Doing it well . . . buyer."* James Newcomb, "The Future of Energy Efficiency Services in a Competitive Environment," Strategic Issues Paper, E Source, 1994, p. 17.

78 *The beauty in good design*: See Gelernter.

79 *Paul Klee said*: The quotation and remark about Klee are from Annie Dillard, *The Writing Life,* Harper & Row, New York, 1989.

84 *digital architecture . . . surfaces*: Paul Goldberger, "Digital Dreams," *The New Yorker*, March 12, 2001.

Chapter 5: Engineering and Its Solutions

90 *Standard Engineering*: Thomas Kuhn called routine science "normal science," and Edward Constant, following him, talks of *normal* engineering. I do not like the term "normal" because it implies parallels to activities in science that may be not be there. I prefer to call this *standard* engineering.

92 *Boeing 747 in . . . 1960s*: The quotes on the 747 are from Peter Gilchrist, *Boeing 747*, 3rd Ed, Ian Allen Publishing, Shepperton, UK, 1999; and Guy Norris and Mark Wagner, *Boeing 747: Design and Evolution since 1969*, MBI Publishing Co., Osceola, WI, 1997; also, personal communication, Joseph Sutter, Boeing, November 2008.

93 *Getting things to work requires*: Ferguson, p. 37.

98 *envisions the concepts and functionalities*: Ferguson tells us this happens visually. I do not disagree, but I am talking about something that happens on a more unconscious and not necessarily visual level.

98 *Hoare created the Quicksort algorithm*: For more on Quicksort, see Gelernter.

99 *Robert Maillart created a . . . bridges*: Billington.

101 *"Such modifications . . . economies"*: Rosenberg, p. 62.

101 *Mechanisms and Mechanical Devices*: Neil Sclater and Nicholas P. Chironis. *Mechanisms and Mechanical Devices Sourcebook*. 4th Ed. McGraw-Hill, New York, 2007.

102 *Richard Dawkins's memes*: Dawkins, *The Selfish Gene*, Oxford University Press, New York, 1976.

104 *the U.S. Navy, . . . Rickover*: Personal communication from Richard Rhodes. Also, Theodore Rockwell, *The Rickover Effect*, Naval Institute Press, Annapolis, MD, 1992.

105 *The light water . . . dominate*: Robin Cowan, "Nuclear Power Reactors: A Study in Technological Lock-in," *Journal of Economic History* 50, 541–556, 1990; Mark Hertsgaard, *Nuclear Inc: The Men and Money Behind Nuclear Energy*, Pantheon Books, New York, 1983.

106 *The shearmen (who . . . agreement*: The quote is from Malcolm Chase, p. 16, *Early Trade Unionism*, Ashgate, Aldershot, UK, 2000. Inner quote from L. F. Salzmann, *English Industries in the Middle Ages*, pp. 342–343, Oxford University Press, 1923.

Chapter 6: The Origin of Technologies

107 *The Origin of Technologies*: Much of the material of this chapter is taken from my paper, "The Structure of Invention," *Research Policy* 36, 2:274–287, March 2007.

107 *"Add successively as . . . thereby"*: Schumpeter, 1912, p. 64.

111 *"steady aerothermodynamic flow . . . component."*: Quote is from Constant, p. 196. Others besides Whittle and von Ohain had of course experimented with earlier versions of the jet engine.

113 *Whittle, in 1928, mulled*: Constant; Whittle.

113 *John Randall and . . . principle*: Russell Burns, "The Background to the Development of the Cavity Magnetron," in Burns, ed., *Radar Development to 1945*, Peter Peregrenus, London, 1988; E. B. Callick, *Meters to Microwaves: British Development of Active Components for Radar Systems 1937 to 1944*, Peter Peregrinus, London, 1990; and Buderi.

114 *"I asked myself . . . arrangement."*: The Lawrence quote is from his Nobel Lecture, "The Evolution of the Cyclotron," December 11, 1951. Wideröe's paper is "Über ein neues Prinzip zur Herstellung hoher Spannungen," *Archiv für Elektrotechnik*, XXI, 386–405, 1928.

115 *In all these . . . appropriated*: Occasionally principles can be arrived at by systematic investigation of the possibilities. "I therefore started to examine systematically all possible alternative methods," says Francis W. Aston of his explorations that would lead to the mass spectrograph. Aston, "Mass Spectra and Isotopes," Nobel Lecture, December 12, 1922.

115 *"While I was . . . proposals."*: Quoted in Constant, p. 183.

117 *Starkweather solved the modulation problem*: See Gary Starkweather, "High-speed Laser Printing Systems," in M. Ross and F. Aronowitz eds., *Laser Applications* (Vol. 4), Academic Press, New York, 1980; and Starkweather, "Laser Printer Retrospective," in *50th Annual Conference: A Celebration of All Imaging*, IS&T, Cambridge, MA, 1997.

118 *"During a seminar . . . celebrate"*: Townes, p. 66.

119 *In 1928, Alexander Fleming*: On penicillin, see Ronald Hare, *The Birth of Penicillin*, Allen and Unwin, London, 1970; Trevor I. Williams, *Howard Florey: Penicillin and After*, Oxford, London, 1984; Eric Lax, *The Mold in Dr. Florey's Coat*, Henry Holt, New York, 2005; Ronald Clark, *The Life of Ernst Chain: Penicillin and Beyond*, St. Martin's Press, New York, 1985; and Ernst Chain, "Thirty Years of Penicillin Therapy," *Proc. Royal Soc. London, B*, 179, 293–319, 1971.

121 *How exactly does . . . place?*: There is a growing literature on the subconscious processes behind creative insight, for example, Jonathan Schooler and Joseph Melcher, "The Ineffability of Insight," in Steven M. Smith, *et al*, eds., *The Creative Cognition Approach*, MIT Press, Cambridge, MA, 1995.

123 *Molecular resonance was exactly*: See Townes; M. Berlotti, *Masers and Lasers: an Historical Approach*, Hilger, Bristol, 1983; and Buderi.

123 *"It was too . . . sequence."*: Mullis, *Dancing Naked in the Mind Field*. Vintage, New York, 1999.

124 *This wider perspective*: Technology writers call this the combination/accumulation view. Constant uses it beautifully to show how steam turbines, turbo air compressors, and experience with the gas turbine led to the origination of the turbojet.

125 *prior embodiment of the principle*: See Charles Süsskind, "Radar as a Study in Simultaneous Invention," in Blumtritt, Petzold, and Aspray, eds. *Tracking the History of Radar*, IEEE, Piscataway, NJ, 1994, pp. 237–45; Süsskind, "Who Invented Radar?," in Burns, 1988, pp. 506–12; and Manfred Thumm, "Historical German Contributions to Physics and Applications of Electromagnetic Oscillations and Waves," *Proc. Int. Conf. on Progress in Nonlinear Science, Nizhny Novgorod, Russia*, Vol. II; *Frontiers of Nonlin. Physics*, 623–643, 2001.

126 *"There is no . . . 'first.'"*: the quote is from M. R. Williams, "A Preview of Things to Come: Some Remarks on the First Generation of Computers," in *The First Computers—History and Architecture*, Raul Rojas and Ulf Hashagen, eds., MIT Press, Cambridge, MA, 2000.

126 *human interaction and . . . communication*: David Lane and Robert Maxfield talk about *generative relationships* that "can induce changes in the way the participants see their world and act in it and even give rise to new entities, like agents, artifacts, even institutions." Lane and Maxfield, "Foresight, Complexity, and Strategy," in Arthur, Durlauf, and Lane. Aitken (1985, p. 547) says that "[t]o understand the process [of invention] it is essential to understand the previously separate flows of information and stocks of knowledge that came together to produce something new."

129 *says mathematician Kenneth Ribet*: Quoted in Simon Singh, *Fermat's Last Theorem*, Fourth Estate, London, 1997, p. 304.

Chapter 7: Structural Deepening

131 *path of development*: Economists call this a technological trajectory. See Richard Nelson and Sidney Winter, "In Search of a Useful Theory of Innovation," *Research Policy* 6, 36–76, 1977; G. Dosi, "Technological Paradigms and Technological Trajectories," *Research Policy* 11, 146–62, 1982; and Dosi, *Innovation, Organization, and Economic Dynamics*, Edward Elgar, Aldershot, UK, 2000, p. 53. Economists have much to say here that I do not talk about. They examine how development paths are influenced by the knowledge base available to a technology and the science that surrounds it, by incentives that firms face, by how the search process for improvements differs within different bodies of technology, by the patent system and legal environment that surround the technology, by learning effects, and by the structure of the industry the technology fits into.

132 *the new technology . . . specialize*: In biology language, we would say the technology *radiates*.

132 *This is where . . . selection*: For Darwinian approaches see Stanley Metcalfe, *Evolutionary Economics and Creative Destruction*, Routledge, London, 1998; Saviotti and Metcalfe; also Joel Mokyr, "Punctuated Equilibria and Technological Progress," *American Economic Assoc. Papers and Proceedings* 80, 2, 350–54, May 1990; Basalla.

133 *Such obstacles are exasperating*: See Robert Ayres, "Barriers and Breakthroughs: an 'Expanding Frontiers' Model of the Technology-Industry Life Cycle," *Technovation* 7, 87–115, 1988.

133 *a bottleneck that . . . of*: Constant talks of "log-jams and forced inventions" (p. 245), and of "anomaly-induced" technical change (pp. 5, 244).

134 *Structural Deepening*: For an earlier discussion of this, see Arthur, "On the Evolution of Complexity," in *Complexity*, G. Cowan, D. Pines, D. Melzer, eds., Addison-Wesley, Reading, MA, 1994; also Arthur, "Why do Things Become More Complex?," *Scientific American*, May 1993.

139 *In 1955 the . . . wondered*: Frankel, "Obsolescence and Technological Change in a Maturing Economy," *American Economic Review* 45, 3, 296–319, 1955.

139 *"In the situations . . . risk."*: Vaughan, *Uncoupling*, Oxford University Press, New York, 1986, p. 71.

140 *a phenomenon I . . . stretch*: The biological phenomenon of *exaptation* resembles this. This is the use of existing parts for a new purpose: webbing a hand to produce a wing for flight, for instance. Adaptive stretch is slightly different because more often it involves system deepening rather than the commandeering of existing parts for different purposes.

140 *This forced piston engines*: Samuel D. Heron, *History of the Aircraft Piston Engine*, Ethyl Corp., Detroit, 1961; Herschel Smith, *Aircraft Piston Engines*, Sunflower University Press, Manhattan, Kansas, 1986.

141 *resembles the cycle . . . proposed*: Others (e.g., Dosi) have also compared technological trajectories with Kuhn's vision of scientific development.

Chapter 8: Revolutions and Redomainings

145 *They are not invented*: There are occasional exceptions. Programming languages (which are certainly domains) are deliberately put together by individuals or companies.

147 *I am thinking . . . here*: Genetic engineering is wider than my discussion would indicate. It has many agricultural applications, and includes techniques such as the production of monoclonal antibodies. One good account of its early development is Horace F. Judson's "A History of the Science and Technology Behind Gene Mapping and Sequencing," in *The Code of Codes*, Daniel J. Kevles and Leroy Hood, eds., Harvard University Press, Cambridge, MA, 1992.

149 *The economist Carlota Perez*: See Perez. The interpretation of technology buildout in this section is very much my own.

150 *The old dispensations persist*: Edgerton.

150 *A domain morphs*: Rosenberg points out that the original applications of a technology are seldom the ones it ends up with.

152   *the arrival of . . . technology*: It also sets in motion economic growth. How this happens is the subject of a branch of economics called endogenous growth theory. The arrival of a new technology means that the economy needs to use fewer resources on that purpose than it did before, thus releasing them. The knowledge inherent in the new technology can also spill over into other industries. For both these reasons, the economy grows.

152   *The coming of the railroad*: For more detailed accounts of its economic impact, see Robert W. Fogel, *Railroads and American Economic Growth*, Johns Hopkins Press, Baltimore, 1964; Alfred Chandler, *The Railroads*, Harcourt, Brace & World, New York, 1965; and Albert Fishlow, *American Railroads and the Transformation of the Antebellum Economy*, Harvard University Press, Cambridge, MA, 1965.

157   *It persists despite . . . inferiority*: In a slightly different context Hughes talks of "technological momentum," in Smith and Marx.

157   *Paul David gave . . . example*: David, 1990.

159   *It is a . . . knowings*: Michael Polanyi pointed out long ago that much of human knowledge is tacit, indeed that such knowledge is indispensable. "The skill of a driver cannot be replaced by a thorough schooling in the theory of the motor-car." Polanyi, *The Tacit Dimension*, Anchor Books, New York, 1966, p. 20.

160   *it derives collectively . . . beliefs*: "Open innovation," where experts can contribute to innovation via the internet without being physically clustered, would seem to counter this tendency of expertise to cluster. It can be useful and important. But it cannot provide face-to-face interaction, or easily form a culture, or provide a repository of local resources that can be accessed by walking down a corridor. See also Brown and Duguid.

160   *"Whatever was known . . . [Cavendish]"*: Quote is from Brian Cathcart, *The Fly in the Cathedral: How a Group of Cambridge Scientists Won the International Race to Split the Atom*, Farrar, Straus, & Giroux, New York, 2004.

160   *"When an industry . . . material."*: Alfred Marshall, *Principles of Economics*, p. 271, Macmillan, London, 8th Ed., 1890.

162   *In the 1980s . . . industry*: Goodyear stayed in Akron. But after 1990, Akron stopped manufacturing tires for personal vehicles.

162   *All this of . . . competitiveness*: See Johan P. Murmann, *Knowledge and Competitive Advantage*, Cambridge University Press, Cambridge, UK, 2003.

163   *All this is . . . innovation*: For economists' views see Dosi, "Sources, Procedures, and Microeconomic Effects of Innovation," *J. Econ. Literature*, XXVI, 1120–1171, 1988; and R. Nelson and S. Winter, *op. cit.*

Chapter 9: The Mechanisms of Evolution

167   *In the early . . . de Forest*: For a full historical account of early radio, see Aitken, 1976, 1985.

168   *technologies are created . . . technologies*: This does not imply that all components of a novel technology previously exist. The atomic bomb came into existence using methods of separation of the fissionable isotope U235 from the chemically similar U238 isotope that did not exist before. But these were

put together from existing methods: separation by centrifuge, electromagnetic separation, and gaseous-barrier and liquid-thermal diffusion. So when I say that technologies are constructed from ones that previously exist, I mean this as shorthand for saying they are constructed from ones that previously exist or ones that can be constructed at one or two removes from those that previously exist.

171 *In the beginning, . . . harnessed*: See Ian McNeil, "Basic Tools, Devices, and Mechanisms," in *An Encyclopedia of the History of Technology*, McNeil, ed., Routledge, London, 1990.

172 *"The more there . . . curve*: Ogburn, p. 104.

180 *"gales of creative destruction"*: Schumpeter, 1942, pp. 82–85.

181 *An Experiment in Evolution*: Arthur and Polak.

184 *There is a parallel observation in biology*: Richard Lenski, *et al.*, "The evolutionary origin of complex features," *Nature*, 423, 139–143, 2003.

185 *This yielded avalanches*: Polak and I found that these "sand-pile" avalanches of collapse followed a power law, which suggests, technically speaking, that our system of technologies exists at self-organized criticality.

187 *More familiarly, larger . . . combinations*: another important source of this is gene and genome duplication. The Jacob quote is from his *The Possible and the Actual*, Pantheon, New York, 1982, p. 30.

Chapter 10: The Economy Evolving as Its Technologies Evolve

191 *The standard way . . . economy*: The definition is from Dictionary.com. Word Net® 3.0, Princeton University, 2008.

196 *When workable textile . . . arrive*: For accounts of the industrial revolution, see in particular Landes, 1969; Mokyr, 1990; and T. S. Ashton, *The Industrial Revolution*, Oxford University Press, New York, 1968.

197 *"the moral conditions . . . classes"*: Quote is from M. E. Rose, "Social Change and the Industrial Revolution," in *The Economic History of Britain since 1700*, Vol. 1, R. Floud and D. McCloskey, eds., Cambridge University Press, Cambridge, UK, 1981. See also P. W. J. Bartrip and S. B. Burman, *The Wounded Soldiers of Industry*, Clarendon Press, Oxford, UK, 1983.

197 *Labor was much . . . organize*: M. Chase, *Early Trade Unionism*, Ashgate, Aldershot, UK, 2000; and W. H. Fraser, *A History of British Trade Unionism 1700–1998*, Macmillan, London, 1999.

198 *"required and eventually . . . jailer."*: Landes, *op. cit.*, p. 43.

198 *There is . . . inevitable*: The notion that technology determines the future economy and future social relations is called technological determinism. Marx is often accused of it. "The hand-mill gives you society with the feudal lord; the steam-mill society with the industrial capitalist." Rosenberg argues convincingly that Marx was too subtle to be a determinist.

199 *Problems as the . . . Solutions*: Friedrich Rapp mentions this in passing. See his article in Durbin.

199 *But within this . . . disruption*: Schumpeter, 1912.

199 *"goods, the new . . . one."*: Schumpeter, 1942, pp. 82–85.

Notes

Chapter 11: Where Do We Stand with This Creation of Ours?

211 *an evolving, complex system*: See Arthur, 1999.

212 *It is one . . . imperfection*: Says psychologist Robert Johnson, "It seems that it is the purpose of evolution now to replace an image of perfection with the concept of completeness or wholeness. Perfection suggests something all pure, with no blemishes, dark spots or questionable areas. Wholeness includes the darkness but combines it with the light elements into a totality more real and whole than any ideal. This is an awesome task, and the question before us is whether mankind is capable of this effort and growth. Ready or not, we are in that process." R. A. Johnson, *He: Understanding Masculine Psychology*, Harper and Row, New York, 1989, p. 64.

212 *"I like elements . . . function."*: Venturi, p. 16.

213 *"Nature becomes a . . . station"*: Martin Heidegger, *Discourse on Thinking*, John M. Anderson and E. Hans Freund, trans., Harper & Row, 1966.

214 *"Technology is in . . . history."*: William Lovitt, in his introduction to Heidegger, p. xxix.

214 *Technology is a . . . life*: Ellul, 1980, p. 125.

216 *"The Buddha, the . . . flower."*: Robert M. Pirsig, *Zen and the Art of Motorcycle Maintenance*, HarperCollins, New York, 1974.

BIBLIOGRAPHY

Not all of these works are cited in the text, but I have found them all useful. Many other writings are cited in the Notes.

Abbate, Janet. *Inventing the Internet*. MIT Press, Cambridge, MA. 1999.

Aitken, Hugh G. J. *The Continuous Wave*. University Press, Princeton. 1985.

Aitken, Hugh G. J. *Syntony and Spark: The Origins of Modern Radio*. University Press, Princeton. 1976.

Arthur, W. Brian. "Competing Technologies, Increasing Returns, and Lock-In by Small Historical Events," *Economic Journal* 99:116–131. 1989.

Arthur, W. Brian. "Complexity and the Economy." *Science*, 284: 107–109. April 2, 1999.

Arthur, W. Brian. *Increasing Returns and Path Dependence in the Economy*. University of Michigan Press, Ann Arbor. 1994.

Arthur, W. Brian, Steven Durlauf, and David A. Lane, eds. *The Economy as an Evolving Complex System*. Addison-Wesley, Reading, MA. 1997.

Arthur, W. Brian, and Wolfgang Polak. "The Evolution of Technology Within a Simple Computer Model." *Complexity*, 11, 5. 2006.

Baldwin, Carliss Y., and Kim B. Clark. *Design Rules*: The Power of Modularity, Vol. 1. MIT Press, Cambridge, MA. 2000.

Bibliography

Basalla, George. *The Evolution of Technology.* Cambridge University Press, Cambridge, UK. 1988.

Bijker, Wiebe E., Thomas P. Hughes, and Trevor J. Pinch, eds. *The Social Construction of Technological Systems.* MIT Press, Cambridge, MA. 1993.

Billington, David, P. *Robert Maillart's Bridges.* University Press, Princeton. 1979.

Brown, John Seely, and Paul Duguid. *The Social Life of Information.* Harvard Business Press, Cambridge, MA. 2000.

Buderi, Robert. *The Invention that Changed the World* [Radar] Simon & Schuster, New York. 1996.

Bugos, Glenn E. *Engineering the F-4 Phantom II.* Naval Institute Press, Annapolis. 1996.

Campbell-Kelly, Martin, and William Aspray. *Computer: A History of the Information Machine.* Basic Books, New York. 1996.

Castells, Manuel. *The Rise of the Network Society.* Wiley-Blackwell, New York. 1996.

Constant, Edward W. *The Origins of the Turbojet Revolution.* Johns Hopkins University Press, Baltimore. 1980.

David, Paul. "The Dynamo and the Computer." *AEA Papers & Proc.* 80. May 2, 1990.

David, Paul. *Technical Choice, Innovation, and Economic Growth.* Cambridge University Press, Cambridge, UK. 1975.

Dosi, Giovanni, Christopher Freeman, Richard Nelson, Gerald Silverberg, and Luc Soete, eds. *Technical Change and Economic Theory.* Pinter, London. 1988.

Durbin, Paul T., ed. *Philosophy of Technology.* Kluwer Academic Publishers, Norwell, MA. 1989.

Edgerton, David. *The Shock of the Old: Technology and Global History since 1900.* Oxford, New York. 2006.

Ellul, Jacques. *The Technological Society.* Alfred A. Knopf, New York. 1967.

Bibliography

Ellul, Jacques. *The Technological System.* Continuum, New York. 1980.

Ferguson, Eugene. *Engineering and the Mind's Eye.* MIT Press, Cambridge, MA. 1999.

Foster, Richard. *Innovation.* Summit Books, New York. 1986.

Freeman, Christopher. *The Economics of Innovation.* Edward Elgar Publishers, Aldershot, UK. 1990.

Gelernter, David. *Machine Beauty: Elegance and the Heart of Technology.* Basic Books, New York. 1998.

Gilfillan, S. Colum. *Inventing the Ship.* Follett, Chicago. 1935.

Gilfillan, S. Colum. *The Sociology of Invention.* Follett, Chicago. 1935.

Grübler, Arnulf. *Technology and Global Change.* Cambridge University Press, Cambridge, UK. 1998.

Heidegger, Martin. *The Question Concerning Technology.* Harper and Row, New York. 1977.

Hiltzik, Michael A. *Dealers of Lightning: Xerox PARC and the Dawn of the Computer Age.* HarperBusiness, New York. 1999.

Hughes, Thomas P. *Networks of Power: Electrification in Western Society, 1880–1930.* Johns Hopkins University Press, Baltimore. 1983.

Hughes, Thomas P. *Rescuing Prometheus.* Pantheon Books, New York. 1998.

Jewkes, John, David Sawers, and Richard Stillerman. *The Sources of Invention.* Norton, New York. 1969.

Kaempffert, Waldemar. *Invention and Society.* Reading with a Purpose Series, No. 56, American Library Association, Chicago. 1930.

Knox, Macgregor, and Williamson Murray. *The Dynamics of Military Revolution, 1300–2050.* Cambridge University Press, Cambridge, UK. 2001.

Kuhn, Thomas S. *The Structure of Scientific Revolutions.* University of Chicago Press, Chicago. 1962.

Landes, David S. *Revolution in Time.* Belknap Press, Cambridge, MA. 1983.

Bibliography

Landes, David S. *The Unbound Prometheus*. Cambridge University Press, Cambridge, UK. 1969.

MacKenzie, Donald. *Knowing Machines*. MIT Press, Cambridge, MA. 1998.

MacKenzie, Donald, and Judy Wajcman, eds. *The Social Shaping of Technology*. 2nd Ed. Open University Press, Buckingham, UK. 1999.

Martin, Henri-Jean. *The History and Power of Writing*. University of Chicago Press, Chicago. 1988.

McGee, David. "The Early Sociology of Invention." *Technology & Culture* 36:4. 1995.

McGinn, Robert. *Science, Technology, and Society*. Prentice-Hall, New York. 1990.

Mokyr, Joel. *The Gifts of Athena: Historical Origins of the Knowledge Economy*. University Press, Princeton. 2004.

Mokyr, Joel. *The Lever of Riches*. Oxford, New York. 1990.

Ogburn, William F. *Social Change*. 1922. Reprint. Dell, New York. 1966.

Otis, Charles. *Aircraft Gas Turbine Powerplants*. Jeppesen Sanderson Aviation, Englewood, Colorado. 1997.

Perez, Carlota. *Technological Revolutions and Financial Capital*. Edward Elgar, Aldershot, UK. 2002.

Rhodes, Richard. *The Making of the Atomic Bomb*. Simon & Schuster, New York. 1986.

Riordan, Michael, and Lillian Hoddeson. *Crystal Fire: The Invention of the Transistor and the Birth of the Information Age*. W.W. Norton, New York. 1997.

Rogers, G.F.C. *The Nature of the Engineering: A Philosophy of Technology*. Palgrave Macmillan, London. 1983.

Rosenberg, Nathan. *Inside the Black Box: Technology and Economics*. Cambridge University Press, Cambridge, UK. 1982.

Saviotti, P. Paolo, and J. Stanley Metcalfe, eds. *Evolutionary Theories of Economic and Technological Change.* Harwood Academic Publishers, Newark, NJ. 1991.

Schmookler, Jacob. *Invention and Economic Growth.* Harvard University Press, Cambridge. 1966.

Schumpeter, Joseph. *The Theory of Economic Development.* 1912. Reprint. Harvard University Press, Cambridge, MA. 1966. 1934.

Schumpeter, Joseph. *Capitalism, Socialism, and Democracy.* 1942. Reprint. Harper, New York. 1975.

Simon, Herbert. *The Sciences of the Artificial.* MIT Press, Cambridge, MA. 1969.

Simon, John. "From Sand to Circuits: A Survey of the Origins of the Microprocessor," in *From Sand to Circuits,* John J. Simon Jr., ed. Harvard University Press, Cambridge, MA. 1986.

Smith, Merritt, and Leo Marx, eds. *Does Technology Drive History?* MIT Press, Cambridge, MA. 1994.

Stroustrup, Bjarne. *The Design and Evolution of C++.* Addison-Wesley, Reading, MA. 1994.

Susskind, Charles. "The Invention of Computed Tomography," in *History of Technology*: Sixth Annual Volume, A. Rupert Hall, and Norman Smith, eds. Mansell Publishing, London. 1981.

Thurston, Robert. *A History of the Growth of the Steam Engine.* Kegan Paul, Trench, & Co, London. 1883.

Tomayko, James E. *Computers Take Flight: A History of NASA's Pioneering Digital Fly-By-Wire Project.* NASA, Washington, D.C. 2000.

Townes, Charles H. *How the Laser Happened.* Oxford University Press, New York. 1999.

Usher, Abbott Payson. *A History of Mechanical Inventions.* 1929. Reprint. Dover, New York. 1988.

Bibliography

Venturi, Robert. *Complexity and Contradiction in Architecture*. Museum of Modern Art, New York. 1966.

Vincenti, Walter. *What Engineers Know and How They Know It*. Johns Hopkins University Press, Baltimore. 1990.

Waldrop, M. Mitchell. *The Dream Machine: J. C. R. Licklider and the Revolution That Made Computing Personal*. Viking, New York. 2001.

Whittle, Frank. *Jet: The Story of a Pioneer*. Frederick Muller, London. 1953.

Winner, Langdon. *Autonomous Technology*. MIT Press, Cambridge. 1977.

ACKNOWLEDGMENTS

This project has grown, rather haphazardly, over several years. It was initially supported by Ernesto Illy, grew into the Stanislaw Ulam Lectures at the Santa Fe Institute in 1998 and the Cairnes Lectures at the National University of Ireland, Galway, in 2000, and in 2001 began to take shape as a book. I thank my home institutions, the Santa Fe Institute and the Intelligent Systems Lab at PARC, for providing refuge during the research and writing, and my colleagues at both places, in particular, Geoffrey West and Markus Fromherz. The International Institute for Applied Systems Analysis in Austria hosted me as Institute Scholar through parts of the writing; and IBM Almaden provided partial support as a Faculty Fellow. I am grateful to St. John's College library in Santa Fe for allowing me to quietly write there, and to Stanford's Green Library (courtesy of librarian Michael Keller) for access to its stacks. The overall book was supported by a grant from the Alfred P. Sloan Foundation.

I have been lucky to have had as agent John Brockman and as editor Emily Loose. I am grateful to both, and to Emily's team. A number of people read the manuscript and gave me useful feedback. Particularly valuable were Michael Heaney, Henry Lichstein, Jim Newcomb, Kate Parrot, and Jim Spohrer. For technical and definitional advice at various points I thank Giovanni Dosi, Doyne Farmer, Arnulf Grübler, John Holland, Kevin Kelly, Geoffrey Marcy, Nebojsa

Acknowledgments

Nakicenovic, Richard Rhodes, and Peter Schuster. Boeing engineers Mike Trestik and Joseph Sutter commented on the aircraft material. My sons Ronan Arthur and Sean Arthur provided much needed writing criticism. Brid Arthur helped me plan the flow of the book, and Niamh Arthur helped edit the final draft.

One of the joys of the project has been the company of friends and colleagues who have provided intellectual stimulation and moral support over the years. I thank in particular Cormac McCarthy and my SFI co-conspirator David Lane; also Kenneth Arrow, Jim Baker, John Seely Brown, Stuart Kauffman, Bill Miller, Michael Mauboussin, Richard Palmer, Wolfgang Polak, Nathan Rosenberg, Paul Saffo, Martin Shubik, Jan Vasbinder, and Jitendra Singh. Not least, I am deeply grateful to my partner, Runa Bouius, for her patience and support during the time this book was being written.

INDEX

Index

invention, 2, 3, 5, 16, 80, 85, 91, 106–30, 163, 172
  commercial use of, 90, 117
  definition of, 20, 108
  by an individual, 5, 110, 111, 112, 124
  linking of need and effect in, 109–16, 120, 129, 139, 204
  mathematical, 126–29
  mental processes of, 23, 97, 121–23
  novel building blocks of, 129–30
  physical forms of, 116–19, 126
  principles and essence of, 106–24
  scientific, 126–29
  sociology of, 6, 174–76
  synthesis of elements in, 20, 21, 124–26
  see also innovation
iodine-gas cells, 48–49
ions, 25, 63, 80
iron, 58, 74–75, 152, 185

James, Henry, 77
jet engines, 17–20, 34, 35, 38–39, 40, 51–53, 65, 77, 93–96, 107, 113, 115, 133–37, 168, 173

Kelly, Kevin, 28
Keynes, John Maynard, 202
Klee, Paul, 79
klystron tube, 113
knowledge, 27, 59–60, 108, 124, 159–60
  cumulative, 57, 65
  intuitive, 78–79
Kuhn, Thomas, 89, 141–42

labor, 196–98
  division of, 37
  see also factories; trade union
Landes, David, 198
language, 97, 114, 212
  design as expression in, 76–79, 89, 97–98
  domain, 69, 76–80, 147

grammar and vocabulary in, 4–5, 76–78, 79, 102
  programming, 71, 79, 146, 163
  utterances in, 76, 78, 79, 97–98
laser optics, 171
laser printers, 33, 108–9
lasers, 17, 33, 56, 69, 80, 117–18, 171, 174, 177
Lawrence, Ernest, 114–15, 121, 131
Layton, Edwin, 124
Le Corbusier, 99, 212
legal systems, 3, 12, 54, 55, 56, 105, 192, 193, 197, 201, 202
lemurs, 14, 187
levers, 74, 75, 171
Lewis and Clark expeditions, 6
life, 189
  prolongation of, 11
  technology as enhancement of, 216
light, 69, 83
  dark vs., 185, 202
  electric, 150
  from stars, 48–49, 50
limbs, mechanical, 9
logic circuits, 168, 171, 182–85
Los Alamos Historical Museum, 75
lubrication systems, 52, 137
lunar space program, 93

machinery, 1, 16, 75, 139, 157–58, 168, 171, 192, 196, 197, 209
  nature enhanced by, 9, 11–12
  see also specific machines
Macintosh:
  computer, 88–89
  Toolbox, 88
magnetic fields, 58, 59, 61, 83, 113–15, 121
magnetic resonance imaging (MRI), 22, 56, 57, 174
Mahler, Gustav, 54, 56
Maillart, Robert, 99–100, 109
Malthus, Thomas Robert, 127–28
Manhattan Project, 75

Index

ABOUT THE AUTHOR

W. Brian Arthur is a leading thinker in technology, the economy, and complexity science. He has been Morrison Professor of Economics and Population Studies at Stanford, and Citibank Professor at the Santa Fe Institute. He is currently in residence at PARC (formerly Xerox Parc). Arthur is the recipient of the Schumpeter Prize in Economics and the inaugural Lagrange Prize in Complexity Science.